EXPLORING WATER RESOURCES

Michelle K. Hall
Science Education Solutions, Inc.

C. Scott Walker
Oregon State University

Anne K. Huth
Science Education Solutions, Inc.

Larry P. Kendall
Science Education Solutions, Inc.

Jeff S. Jenness
Jenness Enterprises

THOMSON
BROOKS/COLE

Australia · Brazil · Canada · Mexico · Singapore
Spain · United Kingdom · United States

THOMSON
BROOKS/COLE

Exploring Water Resources
Michelle K. Hall
C. Scott Walker
Anne K. Huth
Larry P. Kendall
Jeff S. Jenness

Executive Editor: Peter Adams
Assistant Editor: Carol Benedict
Editorial Assistant: Anna Jarzab
Marketing Manager: Joe Rogove
Marketing Communications Manager: Bryan Vann
Content Project Manager: Belinda Krohmer
Creative Director: Rob Hugel

Art Director: Vernon Boes
Print Buyer: Nora Massuda
Permissions Editor: Roberta Broyer
Cover Designer: Denise Davidson
Cover Image: Digital Vision
Cover Printer: Courier-Kendallville
Printer: Courier-Kendallville

Printed in the United States of America
1 2 3 4 5 6 7 10 09 08 07 06

Library of Congress Control Number: 2006906789

ISBN 0495115126

Thomson Higher Education
10 Davis Drive
Belmont, CA 94002-3098
USA

For more information about our products, contact us at:
Thomson Learning Academic Resource Center
1-800-423-0563

For permission to use material from this text or product, submit a request online at **http://www.thomsonrights.com**.
Any additional questions about permissions can be submitted by e-mail to **thomsonrights@thomson.com**.

Acknowledgments

Series Authors

Michelle K. Hall
Anne K. Huth
C. Scott Walker
Jennifer A. Weeks
Robert F. Butler
Jeff S. Jenness
Larry P. Kendall

Programmers

Jeff S. Jenness
Christian J. Schaller

Media Developers

Christian J. Schaller
Ted Stude

Technical Reviewers

Margaret Chernosky
Lydia Fox
Kathleen Friedman
Robert Kolvoord
Carla McAuliffe
Anne Ortiz
Alan Sills
Jessica Walker

Scientific Reviewers

Robert Butler
Peter Kresan
David Smith
Terry Wallace
Joseph Watkins

Technical Advisors

Tom Garrison
Cheryl Greengrove
Miles Logsdon
David Smith
Jim Washburne
Douglas Yarger

Proofreader

James Kevin

Data Providers

Robert Diaz
Virginia Institute of Marine Science
Don Pool
USGS Tucson Field Office
Pima County Department
of Transportation Technical
Services Division
Nancy Rabelais
*Louisiana University
Marine Consortium*
Christopher Scotese
The Paleomap Project

Field Testers

Tekla Cook
Jo Dodds
Christine Donovan
Lillian Fox
Joshua Hall
Alan Kelley
Bob Kolvoord
Kathy Krucker
Kathy Likos
Daniel Montoya
Stephen Murray
Anthony Ochiuzzi
Graciela Rendon-Coke
Richard Spitzer
Anne Thames
Steven Uyeda
Margaret Wilch

Assistants

Tammy Baldwin
Christine Hallman
Sara McNamara
Marie Renwald
Megan Sayles

The authors would also like to thank the many scientists who took the time to learn about this project and share critical research data and expertise, and to the agencies and individuals that have given us permission to include their outstanding illustrations and photos.

The SAGUARO Project

Michelle K. Hall, Director
Science Education Solutions, Inc.
Los Alamos Research Park
4200 West Jemez Drive
Synergy Center Suite 301
Los Alamos NM 87544-2587

saguaro@scieds.com

http://www.scieds.com/saguaro

Table of contents

Introduction

Getting started iii

    Software needed iii

    What you need to know iii

    Using ArcMap iv

    The ArcMap user interface v

    Basic operations v

    Troubleshooting and support options vi

    Working with large numbers vii

    Estimating percent area viii

Philosophy & design ix

    Philosophy ix

    Instructional Design: The 5-E Learning Cycle x

    Learning science with GIS x

ArcMap QuickReference Sheet xi

Investigations

Unit 1 — Where in the World? 1

    Warm-up 1.1 — Global water sources 3

    Investigation 1.2 — Measuring global water 7

    Reading 1.3 — Utilizing global water reservoirs 15

    Investigation 1.4 — What if the ice sheets melted? 19

    Wrap-up 1.5 — Comparing major reservoirs 29

Unit 2 — The Renewable Resource 33

    Warm-up 2.1 — Too little, too much 35

    Investigation 2.2 — Global precipitation patterns 49

    Reading 2.3 — Moving air and water 57

    Investigation 2.4 — U.S. precipitation patterns 65

    Investigation 2.5 — Surface water flow 79

    Wrap-up 2.6 — The local water picture 85

Unit 3 — Using Water **91**

Warm-up 3.1 — Water in your world 93

Investigation 3.2 — Water for many uses 95

Reading 3.3 — Water at work 107

Investigation 3.4 — Feeding a nation 113

Wrap-up 3.5 — Meeting the challenge 121

Unit 4 — Water for a Desert City **123**

Warm-up 4.1 — Living in a desert 125

Investigation 4.2 — Water in the balance 129

Reading 4.3 — The Tucson Basin aquifer 141

Investigation 4.4 — Groundwater issues 145

Investigation 4.5 — Conserving water 151

Wrap-up 4.6 — The voice of conservation 159

Getting started

Software needed

The investigations in this version of the GIS Investigations for the Earth Sciences series are intended for use on campuses with ArcGIS 9.x laboratory or site licenses. All necessary software and data files should be set up and ready for you to use on campus or lab computers. Your instructor will provide you with information on the location(s) of these computers, and where to access the data files for the investigations.

If you have a valid license for the student edition of ArcGIS 9.x, you may be able obtain the necessary data, additional software, and installation instructions to install and use these materials on your personal computer. Contact your instructor for information on where to obtain the required files.

In addition to ArcGIS 9.x, these materials require the following software. Most are available as free downloads from their respective publishers.

- Saguaro Tools for ArcGIS 9.x (contact your instructor)
- Web browser (Internet Explorer, Netscape, Opera, Mozilla, etc.)
- Windows Media Player
- Google Earth
- Adobe Reader

What you need to know

The authors of this book assume that you have the following computer skills:

- Turning the computer on and, if necessary, logging in as a user.
- Navigating the file system to find folders, applications, and files.
- Launching applications and opening files.
- Opening, closing, moving, and resizing windows.
- Using tools, buttons, menus, and dialog boxes.

Other than these basic computer skills, students and instructors do not need prior experience using ArcGIS software. We do, however, strongly recommend that instructors run through each investigation before presenting it to students, to familiarize themselves with the data and techniques. We also strongly recommend that students and instructors:

- Be familiar with visual cues used in these materials.
- Read all instructions carefully.
- Pay attention to information provided in computer-screen shots and in the left-hand sidebar on each page.

Attention course instructors!

The Instructor Guide (answer key), investigation data, additional software, installation instructions, troubleshooting guide, and other resources can be downloaded from the Instructor Resource Center at:

**http://www.thomsonedu.com/
earthscience/hall/water**

You will need a user name and password, provided by the publisher, to access these files.

Mac OS X compatibility

At this time, ArcGIS 9.x is not compatible with any version of the Macintosh operating system. Users may get satisfactory, but slow, performance running ArcGIS under Virtual PC software on a Macintosh computer, but this is neither recommended nor supported.

What is the difference between ArcGIS, ArcView, and ArcMap?

For all practical purposes, you can think of these as three different names for the same geographic information system software.

Technically, ArcGIS is a family of related tools for managing geographic information systems on a variety of scales, ranging from Web servers to handheld devices.

One of these tools is a package for use on desktop computers, called ArcView. The ArcView 9.x software package consists of three components, each interacting with geographic data in a unique way:

- ArcCatalog — management
- ArcMap — viewing and analysis
- ArcToolbox — modifying data

In these materials, you will only be using the ArcMap component of ArcGIS 9.x.

Using ArcMap

Launching ArcMap and opening project files

- To launch the ArcMap application, click the Start button on the Windows Taskbar and choose **All Programs** > **ArcGIS** > **ArcMap** ( ArcMap).

- If you see the ArcMap dialog box, choose **Browse for file** under the **An existing map** option.

- Choose **File** > **Open**.

- Navigate to the appropriate unit data folder installed on your local hard drive or server and open it. (Your instructor or lab administrator can tell you where to find the investigation folders.)

- Locate the specified ArcMap document file and open it. (The .mxd file extension may or may not be visible, depending on how the computer has been set up.)

Visual cues

Visual cues are used to make the investigation directions easier to follow.

- Text preceded by a computer symbol 🖥 is an instruction—something to do on the computer.

- Names of tools or buttons are capitalized and are followed by a picture of that item as it appears on screen—for example, the Identify tool 🛈.

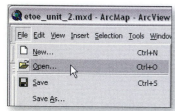
- The > symbol between boldface words or phrases in text indicates a menu choice. Thus, **File** > **Open…** means "pull down the File menu and choose Open…"

Sidebars contain important information!

The page sidebars contain useful information such as definitions, explanations, illustrations, examples, reminders, warnings, tips, and hints. If you are not sure what to do, look for help in the sidebar first.

Closing map files

When you have completed an investigation or must stop for some reason, choose **File** > **Exit** and click **No** when asked if you want to save your changes.

The ArcMap User Interface

Menu Bar

Provides menus for performing various operations.

Title Bar

Shows the name of the current map file.

SAGUARO Tools

Tools provided by the SAGUARO Project for these materials. Note: This toolbar may appear in a different location.

Toolbars

Tools for manipulating and analyzing map data.

Table of Contents

A list of data frames and map layers, and controls for changing the map view.

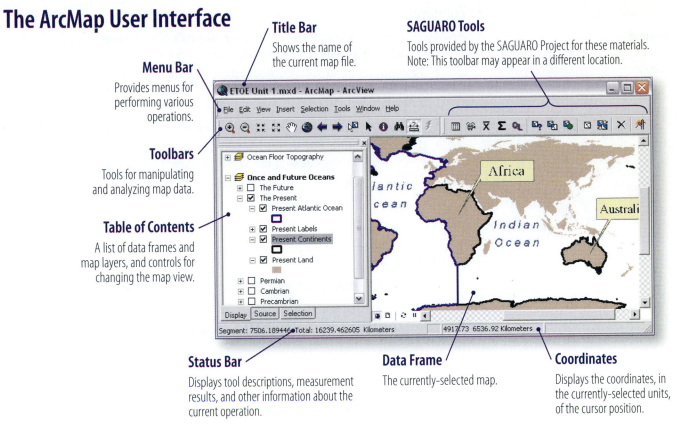

Status Bar

Displays tool descriptions, measurement results, and other information about the current operation.

Data Frame

The currently-selected map.

Coordinates

Displays the coordinates, in the currently-selected units, of the cursor position.

Basic display operations

Activating a data frame

To activate a data frame, right-click its name and choose Activate from the pop-up menu. The title of the activated data frame is highlighted bold.

Expanding and collapsing data frames

To expand a data frame and show its layers, click the expand box ⊞. To collapse a data frame and hide its layers, click the collapse box ⊟.

Selecting layers or layer groups

To select a layer or layer group, click the layer or layer group name. Selected layers or layer groups are highlighted. To select multiple layers or layer groups, hold down the control key while clicking additional names.

Expanding and collapsing layers or layer groups

To expand a layer or layer group, click the expand box ⊞. To collapse a layer or layer group, click the collapse box ⊟.

Turning layers or layer groups on and off

To turn a layer or layer group on, check ☑ the box in front of its name. To turn a layer or layer group off, uncheck the box ☐ in front of its name. If a layer is turned on but is not visible, it may be hidden behind another layer. Try turning off the layers *above* that layer in the Table of Contents.

Table of Contents

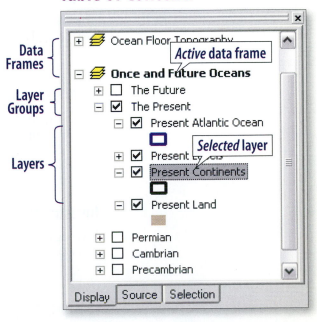

Zooming

The most efficient way to zoom in on a specific area is to drag a box around the area using the Zoom In tool. Drag diagonally from one corner of the area to the opposite corner, then release the button.

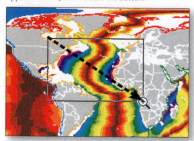

Zooming

ArcMap has tools for zooming—enlarging and reducing areas of the map—that work like the tools you have used in other applications.

- To zoom in on an area, click and drag diagonally with the Zoom In tool to outline the area on the map. When you release the button, the area you selected will rescale to fill the data frame window.
- To zoom out, click anywhere on the map with the Zoom Out tool.
- If you zoom in or out so far that you do not know where you are, undo previous zooms by clicking the Previous Extent button.
- To view the entire data frame, click the Full Extent button.

Want to know more?

If you would like to know more about using ArcGIS, download the **Guide to ArcGIS.pdf** from:

http://www.thomsonedu.com/earthscience/hall/water

Troubleshooting and support options

ArcGIS help

This module provides all of the directions you need to complete the investigations using ArcGIS 9.x. If you have other questions about the capabilities of ArcGIS, choose **Help > ArcGIS Desktop Help**.

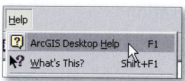

Troubleshooting

The current version of the troubleshooting guide for this series (**Troubleshooting. pdf**) is available from the Instructor Resource Center at:

http://www.thomsonedu.com/earthscience/hall/water

Technical support

If you experience problems with these materials, contact the Thomson Learning technical support team at:

tl.support@thomson.com

1-800-4230563

Investigation data, updates, and resources

Investigation data, updates, and additional resources related to the *GIS Investigations for the Earth Sciences* series can be downloaded from:

http://www.thomsonedu.com/earthscience/hall/water

Working with large numbers

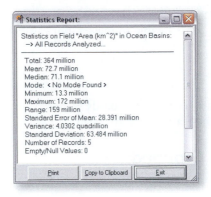

Some of the numbers you will work with in these investigations are quite large. When talking about the amount of water in the ocean or the energy of an earthquake or hurricane, you routinely use values in the billions or even trillions. Where possible, ArcGIS has been modified to make these very large and very small numbers easier to read. For example, in the Statistics Report window shown at left, the total area is given as 364 million, rather than 363958342077361 square kilometers.

Occasionally, you will need to convert millions to billions or thousands, or vice versa. For example, to convert the **Mean** value in the window at left from millions to billions, move the decimal point three places to the *left*. To go from millions to thousands, move the decimal three places to the *right*.

72700 thousand = **72.7 million** = 0.0727 billion

Rounding

Rounding examples

For example, if your number is

319,740,562.85

To round to the nearest ten million:

- Find the ten millions digit (1).
- Look at the number to its right (9). Because it is between 5 and 9, add one to the ten millions digit to make it 2.
- Change the whole numbers to the right of the ten millions digit to zeros and drop the decimal point and everything to its right. The result is **320,000,000**.

Rounding to the nearest...

...million (1,000,000) = **320,000,000**
   (adding 1 to 319 gives 320)
...hundred thousand (100,000) = **319,700,000**
...ten thousand (10,000) = **319,740,000**
...thousand (1,000) = **319,741,000**
...hundred (100) = **319,740,600**
...ten (10) = **319,740,560**
...one (1) = **319,740,563**
...tenth (0.1) = **319,740,562.9**

To round to the nearest 0.1 million:

- Find the 0.1 millions digit (7). This is also called the hundred thousands digit.
- Look at the number to its right (4). Because it is between 0 and 4, do not add one to the 0.1 millions digit.
- Insert the decimal point in the proper location. The result is **319.7 million**.

Most of these numbers are approximations, so it does not make sense to be overly precise when you are calculating or recording them. Look at the number written below, and the place value of each of the digits. Face it—when you are talking about nearly 149 billion of something, who cares about hundred-thousandths, or even tens of millions?

Throughout these investigations, you will be asked to round answers to a particular value and number of decimal places, such as "Round your answer to the nearest 0.1 million." Rounding numbers is simple, if you follow these steps. Examples are shown at the left.

- Look only at the numeral to the right of the place value you are rounding to. For example, when rounding to the nearest thousand, look only at the numeral in the hundreds place.

- If the numeral to the right is 0-4, do not change the number you are rounding to. If the number to the right is 5-9, add one to the number you are rounding to.

- Change whole numerals to the right of the place you are rounding to into zeros, and omit all unused decimal places.

- For any number less than 1, include a zero to the left of the decimal point. (Instead of *.79 billion*, write *0.79 billion*.)

Rounding decimal fractions

Rounding decimals works the same way, except that you are rounding to tenths, hundredths, thousandths, and so on. Do not add zeros to the right of the decimal point. In other words, rounding 2.587 to the nearest tenth is 2.6, *not* 2.600.

Estimating percent area

You will occasionally be asked to estimate the percent area covered by land, ocean, or some other feature. This is a difficult skill for some people to master, but can be learned with practice.

Comparing to standards

One method of estimating coverage is to compare to visual standards. When estimating coverage you need to consider how the features are arranged.

Cloud cover exercise

Here is a simple activity that demonstrates the confusing nature of cover estimates.

- Take two full sheets of blue paper and one of white paper. The blue paper represents sky, and the white paper represents clouds.
- Cut the white sheet in half. Tear or cut the first half of the white sheet into large pieces and glue them onto one of the blue sheets without overlapping.
- Repeat the step above with the other half of the white sheet and the other blue sheet. This time, cut or tear the white sheet into small chunks before gluing them on.

In both cases, the cloud cover is 50 percent. Half of the blue sky is covered by white clouds, but the sheet covered by large clouds appears more open than the sheet covered by small clouds.

Random		Grouped
	0%	
	10%	
	20%	
	30%	
	40%	
	50%	
	60%	
	70%	
	80%	
	90%	
	100%	

Gridding

Another approach to estimating coverage is to divide the area up into a grid, either mentally or physically, and determine the number of grid squares that are at least half-covered. To find the percent coverage, calculate the ratio of covered squares to total squares and multiply by 100.

In the example at left, approximately 20 of the 50 squares are at least half covered.

20/50 × 100 = 40% coverage

Philosophy & design

Philosophy

Thinking scientifically

An Earth scientist makes a living by observing and measuring nature. Whether recording and analyzing earthquakes or measuring subtle changes in sea surface temperature over many decades, a successful Earth scientist relies heavily on his or her ability to recognize patterns. Patterns in space and time are the keys to many of the great discoveries about how Earth works. The investigations in this series are designed to help you develop your ability to recognize and interpret nature's fundamental patterns by exploring recent scientific data using a computer and geographic information system (GIS) software.

Most of these patterns are presented through maps, which are among scientists' most important tools. Maps allow you to visually explore spatial relationships between phenomena such as surface winds and ocean currents; natural features such as continents and ocean basins; and human features such as countries and cities. Behind each map layer is a table containing an extensive database of information about each feature in that layer. By carefully analyzing these data, you can identify patterns in the data that are difficult to discover through visual examination alone.

Planning to learn

Each unit of the series leads you through a well-tested learning process that builds upon your existing knowledge. Each unit begins with a warm-up exercise designed to stimulate your thinking about the major concepts presented in the unit and the key questions that motivate and guide scientific research. It will help you frame your own questions about the topic—questions that you may be able to answer for yourself as you learn more in later investigations.

In the first investigation, you will explore maps and data looking for patterns. As you examine these patterns, you should ask yourself questions such as:

- Where do they occur? (or not occur?)
- Why does this pattern occur here and not elsewhere?
- What might cause this pattern?
- What else is spatially associated with this phenomenon?
- Do these things usually occur together in the same places?
- How has this pattern changed spatially through time?

A brief reading provides key background information about scientific principles and concepts, and should help you begin to answer the questions raised earlier.

Finally, in one or more additional investigations, you apply your new knowledge to solve a particular problem. This helps you measure your understanding of the material and apply the concepts you have learned to a new location or situation.

What is a GIS?

GIS provides tools for organizing, manipulating, analyzing, and visualizing information about the world using digital maps and databases.

GIS made easier

The purpose of these investigations is not simply to learn how to use GIS, but to use one as a tool to explore and learn about natural processes and features and how they relate to humans and human activities. For this reason, all of the data have been assembled into ready-to-use projects, and complex operations have been eliminated or simplified. Although it is helpful for you to have basic computer skills, you do not need experience with ArcGIS software to complete the investigations. The ArcMap user interface has been modified to simplify complex and repetitive processes. Directions for each task are provided in the text, so you will learn to use the tool as you explore with it. The investigations barely scratch the surface of the data that have been provided, and we encourage you to explore the data on your own.

Instructional Design: The 5-E Learning Cycle

This series was designed using the 5-E learning cycle model, which promotes inquiry and exploration as a process for learning science. The Learning Cycle, originally credited to Karplus and Thier (*The Science Teacher*, 1967) and later modified by Roger Bybee for the Biological Sciences Curriculum Study (BSCS) project, proposes that learning something new or understanding something familiar in greater depth involves making sense of both prior experience and firsthand knowledge gained from new explorations. The 5-E model divides learning experiences into five stages: Engage, Explore, Explain, Elaborate, and Evaluate. Each stage builds on the previous stages as you construct new understanding and develop new skills.

Learning science with GIS

Geographic Information Systems (GIS) provide an ideal vehicle for learning topics in Earth and environmental sciences and helping you develop scientific problem-solving skills. Formerly limited to professionals with access to high-end computer workstations, today GIS is accessible to many, and is being used by students from elementary through graduate school. GIS has a number of advantages over traditional materials when used as an instructional tool. These include:

- **Data visualization**—GIS-based investigations allow you to identify and characterize relationships by manipulating multiple visual representations of data, including dynamic and customizable maps, tables, charts, and animations.

- **Data analysis**—Analytical tools enable you to quantify relationships within and among spatial data sets using database functions, statistical analyses, and spatial overlay operations.

- **Multimedia integration**—Other forms of digital information, including animations, video, audio, and digital stills, can be woven into GIS activities, greatly enriching and extending your learning experience.

- **Technology literacy and transferable skills**—The use of GIS promotes technology literacy and provides you with skills transferable to your own research, other course work, and the workplace.

GIS-based instructional materials have the potential to enhance your learning by reinforcing concepts through discovery and by improving problem solving, visualization, and computational skills.

The 5-E Learning Cycle

Engage (Warm-Up)

This stage is designed to help you understand the learning task and make connections to past and present learning experiences. It should stimulate your interest and prompt you to ask your own questions about the topic.

Explore (Investigation)

In this stage, you investigate key concepts by exploring scientific, geographic, and economic data sets. You begin identifying patterns in the data and connecting them to Earth processes. This further stimulates curiosity and new questions develop. You may diverge from the written investigation to explore your own questions, continually building on your knowledge base. Through this process of questioning and exploration, you begin to formulate your understanding of basic concepts.

Explain (Reading)

This stage introduces you more formally to the lesson's scientific and geographic concepts. You should gain a better understanding of major concepts, acquire important terminology, and verify answers to questions or problems posed earlier. In addition, more abstract concepts not easily explored in earlier activities are introduced and explained.

Elaborate (Investigation)

In the Elaborate stage, you will expand on what you have learned and apply your newfound knowledge to different situations. You will test ideas more thoroughly and explore deeper relationships.

Evaluate (Wrap-up)

At the end of each unit, you will use your understanding of key concepts to propose explanations and solutions to local or regional problems.

GIS Investigations for the Earth Sciences

Quick Reference Sheet

The ArcMap User Interface

Menu Bar

Provides menus for performing various operations.

Toolbars

Tools for manipulating and analyzing map data.

Table of Contents

A list of data frames and map layers, and controls for changing the map view.

Status Bar

Displays tool descriptions, measurement results, and other information about the current operation.

Title Bar

Shows the name of the current map file.

SAGUARO Tools

Tools provided by the SAGUARO Project for these materials. Note: This toolbar may appear in a different location.

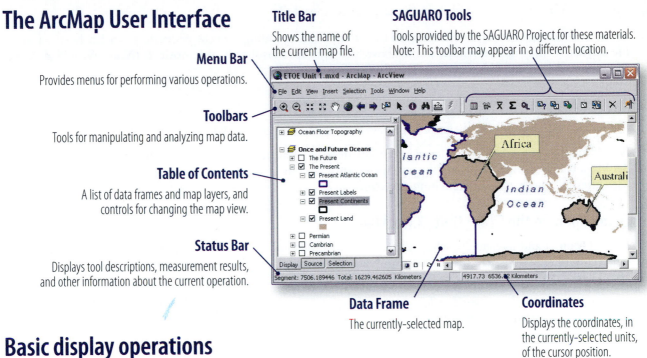

Data Frame

The currently-selected map.

Coordinates

Displays the coordinates, in the currently-selected units, of the cursor position.

Basic display operations

Activating a data frame

To activate a data frame, right-click its name and choose Activate from the pop-up menu. The title of the activated data frame is highlighted bold.

Expanding and collapsing data frames

To expand a data frame and show its layers, click the expand box ⊞. To collapse a data frame and hide its layers, click the collapse box ⊟.

Selecting layers or layer groups

To select a layer or layer group, click the layer or layer group name. Selected layers or layer groups are highlighted. To select multiple layers or layer groups, hold down the control key while clicking additional names.

Expanding and collapsing layers or layer groups

To expand a layer or layer group, click the expand box ⊞. To collapse a layer or layer group, click the collapse box ⊟.

Turning layers or layer groups on and off

To turn a layer or layer group on, check ☑ the box in front of its name. To turn a layer or layer group off, uncheck the box ☐ in front of its name. If a layer is turned on but is not visible, it may be hidden behind another layer. Try turning off the layers *above* that layer in the Table of Contents.

Table of Contents

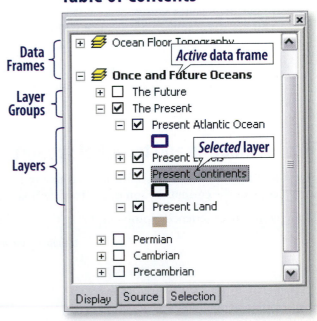

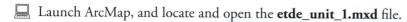

Opening ArcMap document (.mxd) files

At the beginning of each investigation, you will launch the ArcMap application and use it to open an ArcMap document file.

The instruction will look like this:

 Launch ArcMap, and locate and open the **etde_unit_1.mxd** file.

 ArcMap document files end with an .mxd file extension. Depending on how your computer has been set up, file extensions may not be visible, so the above file may appear as simply **etde_unit_1**. (Note: We will also refer to ArcMap document files simply as *map files*.)

Launching ArcMap

To launch the ArcMap application, click the Windows Start menu and choose **All Programs** > **ArcGIS** > **ArcMap**.

Note: If there is an ArcMap shortcut icon on the desktop, Start menu, or taskbar, you can also use it to launch ArcMap. It may or may not show the .exe file extension. Your instructor or lab administrator may provide you with alternate directions for launching ArcMap.

Locating and opening the EXXX_Unit_X.mxd file

Choose **File** > **Open…** and navigate to where the specified ArcMap document file is located. Your instructor or lab administrator should tell you where the investigation data files are located. They will be in folders named with the module abbreviation and unit number, such as ETDE_Unit_1. Remember, you may not see the .mxd file extension.

Important note: Do not call or e-mail technical support if you cannot find an ArcMap document file. Your instructor or lab administrator can tell you where to find it!

Zooming

ArcMap has tools for zooming—enlarging and reducing areas of the map—that work like the tools you have used in other applications.

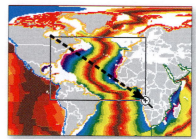

- To zoom in on an area, click and drag diagonally with the Zoom In tool to outline the area on the map. When you release the button, the area you selected will rescale to fill the data frame window.

- To zoom out, click anywhere on the map with the Zoom Out tool.

- If you zoom in or out so far that you do not know where you are, undo previous zooms by clicking the Previous Extent button.

- To view the entire data frame, click the Full Extent button.

Troubleshooting and technical support

Your instructor or lab administrator may be able to help with common problems. If necessary, they can provide you with a copy of the current version of the troubleshooting guide for this series, **Troubleshooting.pdf**.

If you continue experiencing problems using these materials, contact the Thomson Learning technical support team at:

tl.support@thomson.com *or* **1-800-423-0563**

Unit 1
Where in the World?

In this unit, you will

- *Identify global water reservoirs.*

- *Explore challenges to providing safe freshwater to the world's population.*

- *Estimate the volume of water in the oceans, ice sheets, and atmosphere.*

- *Examine the hydrologic cycle, residence time of water, and challenges of recovering water from each global reservoir.*

- *Evaluate the impact of global sea level changes on society.*

Earth is often called the "water planet." With three fourths of its surface covered by water in some form, and being the only planet with liquid water on the surface, this is a fitting description.

Warm-up 1.1

Global water sources

To live, humans need three things: air, water, and food. Except in unusual situations, we can survive 4–6 minutes without air, 3–5 days without water, or up to 3–6 weeks without food. Earth has plenty of air, though its quality is sometimes a concern. We also have abundant water resources—over three fourths of Earth's surface is covered by water in some form.

The water planet

Slightly over three fourths (75%) of Earth is covered by water in either liquid or solid form.

The health and economic welfare of the global population depends on a steady supply of fresh, uncontaminated water, so developing ways to extract usable water from different reservoirs is very important. Many existing reservoirs are being depleted by withdrawals and stressed by contamination. Although future management of our water resources will require significant conservation efforts, we may also find ways to make unusable water usable, or to harvest freshwater more economically from reservoirs previously considered impractical. In this activity, you will identify and determine the size and importance of global water reservoirs and think about how your community may utilize these reservoirs.

1. Look at the image of Earth on the left. What are the three different forms, sometimes called *states*, of water represented in the picture?

    a.

    b.

    c.

Earth's storage areas for water, some seen and some unseen, are called global *reservoirs*. These reservoirs are found above, on, and below Earth's surface.

2. Why is water important? How is it used in society?

Need a hint?

Don't forget that water can be stored in one of three physical states.

3. The ocean is Earth's largest reservoir of water. List as many of Earth's other water reservoirs as you can in Table 1.

Table 1— Global water reservoir predictions

Global reservoir	Size rank	Accessibility	Why?
oceans	1		

4. Rank the reservoirs you listed in Table 1, from largest to smallest based on your best guess. Use 1 to represent the largest reservoir, 2 for the next largest, and so on.

5. Classify the reservoirs you listed in Table 1, on your estimate of their accessibility, or how easily freshwater could be obtained from them, as either *easy, moderate,* or *difficult.*

6. In the **Why?** column of Table 1, briefly explain the reason for your accessibility classification.

Only pure freshwater can be used for drinking, but most of Earth's water is not suitable for drinking. We call suitable water "potable" and unsuitable water "nonpotable."

7. List three uses for nonpotable water.

a.

b.

c.

8. Why is ocean water generally considered unusable for drinking?

Water in your community: A computer lab or library activity

Communities face three major issues in their efforts to supply water to their citizens:

- Quantity — Is there enough water?
- Quality — Is it clean and safe for its intended purpose?
- Accessibility — Is it easy and economical to obtain and distribute?

Conduct your own research to find out where your community's water comes from, and the challenges your community faces in providing a clean and abundant water supply. Be prepared to report on your findings, along with any references you used in your research. You may find this information in your school or public library, on a community Web page, or by calling your local water company.

9. Where does your community's drinking water come from?

10. Was this one of the reservoirs you listed in Table 1? If not, why do you think you missed this reservoir?

11. What challenges does your community face in providing a safe and adequate water supply?

12. How is your community working to ensure a safe and adequate water supply?

Investigation 1.2

Volume: a measure of water

For liquids such as water, volume is a way of describing the amount of the substance, typically in units of cubic feet or cubic meters. For example, drinking water is often sold in one-gallon containers.

Four oceans or five?

Historically, maps showed four major oceans — the Atlantic, Pacific, Indian, and Arctic Oceans. Modern maps often show a fifth ocean, the Southern Ocean, encircling Earth just north of Antarctica.

Bathymetry — the measuring of water depth. The root, *bathy*, comes from a Greek word meaning *deep*. The suffix *-metry* comes from another Greek word meaning *to measure*.

Measuring global water

Water is essential for life as we know it. It exists on Earth in three forms — solid (ice and snow), liquid, and gas (water vapor). Oceans, lakes, and rivers are the most obvious places to find water, but it is also in the air as invisible water vapor and as clouds. Water is also found in the soil, in water-bearing rock formations or *aquifers*, and in frozen soil or *permafrost*. Even the driest deserts show signs of water including dried-up river channels and lake beds. Water is found in many places, yet how much of it is easily available for our use? In this investigation, you will explore this question by examining three global water reservoirs in detail.

Earth's oceans

Most of Earth's water is in its oceans. To begin your investigation, you will estimate the *volume* of water the oceans contain.

🖥 Launch ArcMap, then locate and open the **ewr_unit_1.mxd** file.

Refer to the tear-out Quick Reference Sheet located in the Introduction to this module for GIS definitions and instructions on how to perform tasks.

🖥 In the Table of Contents, right-click the **Earth's Oceans** data frame and choose Activate.

🖥 Expand the **Earth's Oceans** data frame.

This data frame shows Earth's five ocean basins, each outlined in a different color. Ocean depths, determined by bathymetry, are represented by five layers: **Atlantic Depth**, **Indian Depth**, **Pacific Depth**, **Arctic Depth**, and **Southern Depth**. Study the legend for the ocean depth classifications, shown below the **Southern Depth** layer in the Table of Contents.

These ocean-depth layers contain a lot of data, so they take a long time to redraw on your screen. It is a good idea to display only the layers you need at a given time. For now, however, leave all five turned on.

1. How deep are the deepest parts of the oceans? Use the legend to determine the range of greatest depths. Round to the nearest 0.1 km and include units of measurement in your answer.

To help you visualize the shape of the ocean basins, you will examine cross sections of the basins, called *bathymetric profiles*.

🖥 Click the Media Viewer button 🎞.

🖥 Choose the **Bathymetric Profile Movie**. Click **OK**.

🖥 Close the Media Viewer window when you are finished viewing the movie.

🖥 Turn on the **Bathymetric Profiles** layer.

🖥 Select the **Bathymetric Profiles** layer.

This layer shows 12 line segments, called *transects*, drawn across different regions of the world. Each transect is linked to a profile that shows the ocean depth along that transect.

💻 Use the Hyperlink tool ⚡ to click on a transect.

An image window will open, showing the profile of the ocean floor along that line segment. If an image window does not open, make sure the **Bathymetric Profiles** layer is selected.

Negative elevation?

The y-axis in each bathymetric profile expresses depth as a negative elevation, or depth *below* sea level.

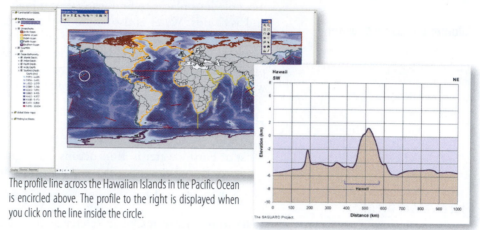

The profile line across the Hawaiian Islands in the Pacific Ocean is encircled above. The profile to the right is displayed when you click on the line inside the circle.

After viewing all the bathymetric profiles, and using the legend on the map, answer the following questions.

2. Where are the ocean basins deepest? Describe your observations from the bathymetric profiles.

3. Is the average or *mean* depth of each ocean approximately the same? Explain.

💻 Close the profile windows when you are finished.

💻 Turn off all layers except for the **Countries** and **Atlantic Depth** layers.

Earlier, you estimated the depths of the ocean basins by eye, using the ocean depth legend. Next you will determine the maximum and mean depth of each of the five oceans more precisely, beginning with the Atlantic Ocean.

How to calculate statistics

💻 Click the Statistics button ⊠.

💻 In the Statistics window, calculate statistics for **all features** of the **Atlantic Depth** layer, using the **Depth (km)** field.

💻 Click **OK**. Be patient while the statistics are calculated.

4. Record the **Maximum** and **Mean** depths of the Atlantic Ocean in Table 1 on the following page. Round both to the nearest 0.1 km.

Table 1 — Ocean volume

Ocean	Maximum depth *km*	Mean depth *km*	Surface area *km²*	Volume *km³*
Atlantic				
Indian				
Pacific				
Arctic				
Southern				
Total Surface Area and Volume				

🖥 Close the Statistics window.

🖥 Repeat this process to calculate and record the maximum and mean depths for each of the four remaining oceans in Table 1. (Note: Remember to turn on and select the appropriate ocean basin layer each time you calculate the statistics. Be sure to round the depths to the nearest 0.1 km.)

🖥 Turn off all five of the ocean depth layers.

Next you will estimate the volume of each ocean by multiplying the ocean's surface area by its mean depth.

🖥 Turn on the **Ocean Basins** layer.

🖥 Click the Identify tool 🛈.

🖥 In the Identify Results window, select the **Ocean Basins** layer from the drop-down list of layers.

🖥 Click on the Atlantic Ocean to display information about the Atlantic Ocean basin.

5. Round the surface area [**Area (km^2)**] of the Atlantic Ocean to the nearest 100,000 km² and record it in Table 1.

🖥 Repeat this process to record the surface area for each of the four remaining oceans in Table 1. (Note: Remember to round each area to the nearest 100,000 km².)

🖥 Close the Identify Results window when you are finished.

6. Calculate the volume of each ocean (see sidebar for help), and record the volume for each ocean in Table 1. Round to the nearest 1,000,000 km³.

7. Add the surface areas of the five oceans and record the total ocean surface area in Table 1.

8. Add the volumes of the five oceans and record the total ocean volume in Table 1.

Calculating volume

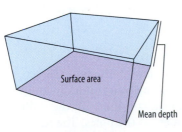

You can calculate the volume of a liquid by multiplying its surface area by its mean depth.

volume = area × mean depth

Water frozen in continental ice sheets

Another important water reservoir is Earth's ice sheets. Land masses near the North and South Poles stay cold enough throughout the year that what little precipitation occurs there falls as snow. Over millions of years, this snow has compacted into a solid, permanent ice sheet a thousand or more meters thick. Next, you will estimate the volume of Earth's freshwater that is locked up in these ice sheets.

- 🖥 Click the QuickLoad button.

- 🖥 Select **Data Frames**, choose **Continental Ice Sheets** from the list, and click **OK**.

Location map

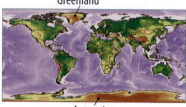

This data frame shows Earth's two large continental ice sheets, covering Antarctica and Greenland.

First, you will find the mean thickness of the Antarctic ice sheet.

- 🖥 Select the **Antarctica Ice Thickness** layer.

- 🖥 Click the Statistics button.

- 🖥 In the Statistics window, calculate statistics for **all features** of the **Antarctica Ice Thickness** layer, using the **Ice Thickness (km)** field.

- 🖥 Click **OK**. Be patient while the statistics are calculated.

The mean thickness of Antarctic ice is reported as the **Mean**.

9. Round the mean thickness to the nearest 0.1 km and record it in Table 2.

Table 2 — Ice sheet volume

Ice sheet	Mean thickness *to nearest 0.1 km*	Surface area *to nearest 100,000 km²*	Ice volume (mean thickness x surface area) *to nearest 100,000 km³*	Water volume (ice volume x 0.9) *to nearest 100,000 km³*
Antarctica				
Greenland				
			Total water volume	

- 🖥 Close the Statistics window.

- 🖥 Select the **Greenland Ice Thickness** layer.

- 🖥 Repeat the statistics procedure above to find the mean thickness of the Greenland ice sheet and record it in Table 2.

- 🖥 Close the Statistics window.

Next you will find the surface area of each ice sheet by viewing them in Google Earth. There are disadvantages to viewing Earth as a flat map. It is not obvious that a feature that disappears off the left edge of the map continues on the right edge, or that features at the top or bottom actually meet at the poles. The real Earth is nearly a sphere, and viewing Earth as a globe will enable you to see Antarctica and Greenland in their entirety.

- 🖥 Launch the Google Earth application.

- 🖥 Double-click the **Antarctica Ice Sheet Thickness.kmz** file located in the **ewr_unit_1** folder. This file should open within the Google Earth application.

Now you are viewing Earth as a globe, looking directly at Antarctica. Use the **Antarctica Ice Thickness** legend in ArcMap to examine the ice thickness in Google Earth.

🖥 Left-click and drag the cursor on the globe to turn it in any direction.

🖥 Click on the text "**Antarctica Ice Sheet Thickness**" in the Google Earth legend on the left side of the map.

A text box should appear on the map, showing the surface area of the Antarctica ice sheet. The surface area has been rounded to the nearest 100,000 km².

10. Record the surface area of Antarctica in Table 2 on the previous page.

🖥 Close the **Antarctica Ice Sheet Thickness.kmz** file and do not save changes.

🖥 Double-click the **Greenland Ice Sheet Thickness.kmz** file located in the **ewr_unit_1** folder. This file should open within the Google Earth application.

Now you are looking directly at Greenland on the globe. Use the **Greenland Ice Thickness** legend in ArcMap to examine the ice thickness in Google Earth.

🖥 Click on the text "**Greenland Ice Sheet Thickness**" in the Google Earth legend on the left side of the map.

A text box should appear on the map, showing the surface area of the Greenland ice sheet. The surface area has been rounded to the nearest 100,000 km².

11. Round the surface area of Greenland to the nearest 100,000 km², and record it in Table 2.

🖥 Quit Google Earth, do not save changes, and return to ArcMap when you are finished.

12. Calculate the volume of each ice sheet, round to the nearest 100,000 km³, and record it in the **Ice volume** column of Table 2. (Hint: volume = mean thickness × surface area.)

Liquid water occupies a smaller volume than solid water (ice). If the ice sheets melted, the resulting liquid water would occupy only 90 percent of the original volume of the ice.

13. Calculate the volume of liquid water in the ice sheets by multiplying the volume of each ice sheet by 90 percent (0.9). Round to the nearest 100,000 km³, and record the results in the **Water volume** column of Table 2.

14. Add the water volumes of the two continental ice sheets, and record it in the box labeled **Total water volume** in Table 2.

15. Why do you think the freshwater in the ice sheets is generally considered unavailable for human use?

Polar deserts?

Although covered in ice, Antarctica is a very dry continent because its mean annual precipitation is only about 5 cm (2 in) — about the same as the Sahara Desert. The main difference is that water in the Sahara evaporates rapidly, whereas ice in Antarctica returns to the atmosphere very slowly.

Other ice on Earth

In addition to the continental ice sheets that cover Greenland and Antarctica, large areas of Earth are covered by glaciers and sea ice.

Water vapor in the atmosphere

The third global water reservoir you will examine is the atmosphere. Clouds in the atmosphere are made up of ice crystals and water droplets, but the air itself also contains water in the form of invisible water vapor.

🖥 Click the QuickLoad button 🔲.

🖥 Select **Data Frames**, choose **Global Water Vapor** from the list, and click **OK**.

🖥 Select the **Mean Water Vapor** layer.

This data frame shows the mean amount of water vapor in the atmosphere during a year. The **Mean Water Vapor** layer shows the amount of water vapor in a column of air reaching from Earth's surface to the top of the atmosphere, a height of about 50 km. Examine this layer's legend and then study the map.

16. Briefly describe how water vapor is distributed in the atmosphere. For example, where do you find the most water vapor and where do you find the least?

A "column" of air?

air column

You can picture a column of air as a tall, transparent building reaching from Earth's surface to the top of the atmosphere.

🖥 Turn on the **Mean Surface Temperature** layer.

The **Mean Surface Temperature** layer shows the mean temperatures of the Earth's surface.

17. Where is the mean temperature highest?

18. How does the mean temperature relate to the amount of water vapor in the atmosphere? Explain.

To determine the total volume of water vapor in the atmosphere, you will simply add the volumes of the individual air columns together.

🖥 Click the Statistics button 🔲.

🖥 In the Statistics window, calculate statistics for **all features** of the **Mean Water Vapor** layer, using the **Water Vapor Volume (km3)** field.

🖥 Click **OK**. Be patient while the statistics are calculated.

The total water vapor is given as the **Total.**

19. How much water vapor is stored in Earth's atmosphere? Include proper units and round to the nearest 1,000 km³.

20. How does water leave this atmospheric reservoir?

21. Describe the challenges you think we face in extracting usable amounts of freshwater directly from the atmosphere.

🖥 Close the Statistics window.

🖥 Quit ArcMap and do not save changes.

Earth's total water budget

Earth's water is stored in places other than the oceans, continental ice sheets, and the atmosphere. To get an overall picture of the global water budget, you need to consider other reservoirs. Table 3 below lists Earth's water reservoirs. The volumes of the reservoirs you have not already examined are provided for you in the table.

22. Fill in the **Water volume (km³)** column for the oceans (from Table 1), continental ice sheets (from Table 2), and the atmosphere (from question 19) in Table 3.

Table 3 — Total volume of water on Earth

Reservoir	Water volume km^3	Percent of total water
Groundwater	8,340,000	0.62
Freshwater lakes	125,000	0.0093
Inland seas	104,000	0.0078
Soil moisture	67,000	0.0050
Rivers	1,250	0.000093
Oceans		
Continental ice sheets		
Atmosphere		
Total		100.0*

* Table does not include water contained in Earth's biomass, so the sum will not add up to exactly 100%.

What is biomass?

Biomass refers to living matter and its by-products. These can contain water but are not included in Table 3.

23. Calculate the total volume of water on Earth by adding the individual volumes of each of the reservoirs and record it in the last row in the table. Round to the nearest 1,000,000 km³.

24. Using the total volume of water, calculate the percentage each reservoir contributes to the total, and record it in the **Percent of total water** column in the table. Round to the first two non-zero digits (see sidebar for instructions on calculating percentages).

Calculating percentages

To calculate percentages for Table 3, divide the water volume of each reservoir in Table 3 by the total water volume in Table 3, and multiply by 100. The first few have been done for you.

25. Using the data in Table 3, calculate what percentage of the world's potable water (freshwater that is fit to drink) is directly accessible in surface reservoirs such as freshwater lakes and rivers.

26. Compare the information in Table 3 to the predictions you made about global water reservoirs in Table 1 in Warm-up 1.1. Which reservoirs did you overestimate and which did you underestimate?

Reading 1.3

Utilizing global water reservoirs

The water planet

Imagine that you could travel far out into the solar system, and look back at Earth. What would you see? As this photo of Earth from space shows, you would see little more than a pale blue dot. The size of the dot clearly illustrates how small we are in the overall scheme of things; but it is the color, visible even at this great distance, that reveals the unmistakable presence of water. The abundance of free-flowing, liquid water makes Earth unique in our solar system.

The oceans hold a vast and deep expanse of water, and there is also water in the air we breathe and in the ground beneath our feet. Despite the abundance of water on this planet, over one billion people do not have access to clean, safe water for consumption, sanitation, and hygiene. In developing countries, 90 percent of wastewater is returned untreated into rivers and streams

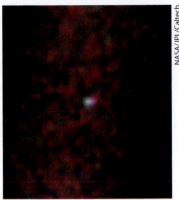

Figure 1. Earth from a distance of 6.5 billion kilometers, as seen by the Voyager 1 spacecraft on February 14, 1990.

that supply water for drinking and hygiene. As a result, over two million people die each year from water-borne diseases including cholera, typhoid, dysentery, infectious hepatitis, and giardiasis. Although the statistics are most alarming for developing countries, all countries face the problem of supplying sufficient amounts of clean, safe water for drinking and other uses. Next, you will look at how water is cycled through Earth's reservoirs, and examine how we use the reservoirs to meet our water needs.

The hydrologic cycle and residence time

Water exists in the atmosphere, oceans, and ice sheets, and on land. It moves among these reservoirs in a process called the **hydrologic cycle**, shown in Figure 2. Water may take many paths among reservoirs and frequently changes its physical state through **evaporation** (liquid to gas), **condensation** (gas to liquid), **freezing** (liquid to solid), **melting** (solid to liquid), and **sublimation** (gas to solid, or the reverse). The movement of water in the hydrologic cycle is driven by solar energy, assisted by gravity. Solar energy causes water to evaporate from oceans, rivers, soils, and even plants (called **evapotranspiration**) and move into the atmosphere. Vertical air currents cause water vapor to rise in the atmosphere, where it cools, condenses, and

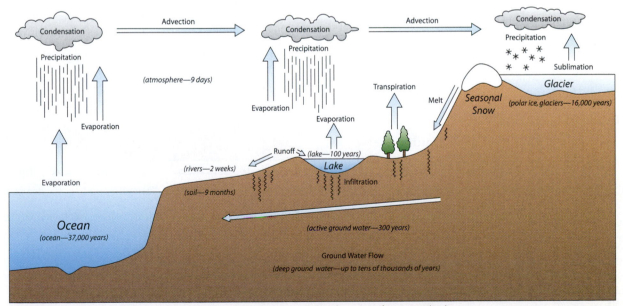

Figure 2. The hydrologic cycle in action. Numbers in parentheses represent typical residence times for water molecules.

returns to Earth's surface as precipitation. When water reaches the surface, gravity draws the water as runoff into lakes, rivers, and streams. Gravity also determines the movement of water through **aquifers** beneath the surface. In addition to the vertical movement of water driven by gravity, the uneven heating of Earth's surface by the sun generates horizontal air movement (wind) in the atmosphere, a process known as **advection.**

Figure 2 (previous page) illustrates the processes that move water among the major global reservoirs in the hydrologic cycle. It also identifies the **residence time** for each reservoir, the average time a water molecule spends in each reservoir. In the atmosphere, the residence time is just nine days. This means that an average water molecule in the atmosphere precipitates out and is replaced by a "new" one every nine days. Residence times in the oceans, ice sheets, and groundwater are much longer, on the order of thousands to tens of thousands of years.

1. The total global population is about 6.5 billion. What percentage of the world's population does not have access to a reliable source of fresh, clean water?

2. In which global reservoir does water reside for the longest time?

3. In which reservoir does water reside for the shortest time?

The ocean as a water source

Oceans contain about 97 percent of Earth's water. Unfortunately, we cannot drink the water from this reservoir or use it to grow food because it is too saline or salty. Water dissolves minerals from the rocks on land and transports them to the oceans. When water evaporates from the oceans, the dissolved minerals are left behind, resulting in the ocean's saline water.

The ocean's salinity level (the amount of dissolved salt) varies, as shown in the global salinity map in Figure 3. The salinity is generally lower in the polar regions where low evaporation rates, combined with meltwater

from glaciers (freshwater), dilute the surrounding ocean waters. Similarly, lower salinity is found along coastlines where rivers empty freshwater into the oceans. In contrast, elevated salinity levels occur where evaporation rates are high and precipitation rates are low, as in the Mediterranean Sea, the Red Sea, and the Gulf of Mexico.

Normal ocean salinity is approximately 35 parts per thousand, which means that a thousand grams of sea water contain about thirty-five grams of salt. The world's oceans, therefore, contain approximately 50 million trillion kilograms of salt, enough to cover the land surface of Earth to a height of around 150 meters.

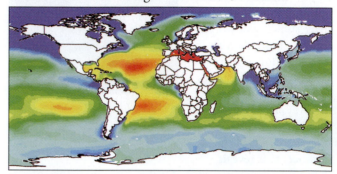

Figure 3. Global ocean surface salinity map. Salinity ranges from below 3.3 percent (dark blue) to above 3.7 percent (dark red). For a complete explanation of why the oceans are salty, see **http://www.palomar.edu/oceanography/salty_ocean.htm**

Why can't we drink sea water?

The most common salt in sea water is sodium chloride (table salt). Sodium is an important nutrient, and is involved in a variety of bodily processes. Normally, the sodium level in a person's bloodstream is in balance (isotonic) with the sodium levels within the individual cells in the body. However, when a person consumes

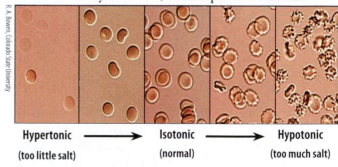

Figure 4. The effect of sodium levels on red blood cells. In **isotonic** blood, the sodium levels of the cells and the surrounding liquid, or plasma, are at equilibrium, and cells appear normal. In **hypertonic** blood, the concentration of sodium in the cells is higher than in the plasma. Water diffuses into the cells, causing them to expand and burst. In **hypotonic** blood, the concentration of sodium in the plasma is higher than in the cells. Water diffuses out of the cells, causing them to collapse.

saltwater, the sodium levels in the bloodstream increase. This difference causes water to move from the (hypotonic) cells into the bloodstream to restore this balance by diluting the excess sodium. The movement of water out of the cells causes them to shrink and stop functioning, a potentially deadly situation. Figure 4 (previous page) illustrates the effect of sodium levels on red blood cells.

Desalination: Making sea water drinkable

Because humans cannot drink saltwater, the salt must be removed from the water to make it suitable for drinking. This process occurs naturally over the oceans, where solar energy evaporates water from the surface of the oceans. As the water evaporates, it leaves the minerals behind. (This explains why all precipitation is freshwater, and why the oceans are salty.) This process of removing salt from ocean water is called **desalination**. Today there are more than 12,000 desalination plants in operation worldwide. Most are located in the Middle East, producing about 19 billion liters of pure water per day. Saudi Arabia alone produces about 3 billion liters of freshwater per day from 27 large desalination plants.

Two basic techniques are used to remove salt from ocean water: **reverse osmosis**, in which the water is pumped through a membrane to filter out salt; and **distillation**, or **evaporation**. Figure 5 illustrates the basic distillation process.

Figure 5. A basic distillation apparatus. Adapted from
http://www.desware.net

Desalination is expensive, due to the amount of power required to pump and filter the water. In California, water produced by desalination plants costs about $2.60 per thousand liters, which seems pretty inexpensive. However, compared to the cost of Colorado River water at less than a penny per thousand liters, it is not surprising that most of the desalination plants are unused, except in emergencies.

Technological improvements may lower the cost of desalinating water. A new desalination plant scheduled to begin operation in Tampa Bay, Florida, is designed to produce water for about 50 cents per thousand liters. Still, we are a long way from turning the world's oceans into an economically viable source of freshwater.

4. What are two processes for desalinating water?

5. Why is it so expensive to desalinate water?

Ice sheets as a water source

As the largest reservoir of freshwater on Earth, the continental ice sheets have long been considered a potential source of drinking water. Every year about 40,000 icebergs break off the Greenland ice sheet. Some of these are pushed southward by the Labrador Current into the North Atlantic Ocean. Because the average Arctic iceberg is about the size of a 15-story building and contains the equivalent of about 114 million liters of water, icebergs seem like a simple solution to water shortages. Antarctic icebergs, like the one shown in Figure 6, are even larger.

Figure 6. A massive iceberg known as B-15 broke off Antarctica's Ross Ice Shelf in March 2000. Among the largest ever observed, the iceberg is about 11,000 square kilometers — nearly the size of the state of Connecticut.

Schemes to lasso and tow icebergs have been discussed for years, but have generally been dismissed due to economic and environmental considerations. Recently, an Australian scientist has proposed wrapping Antarctic icebergs in plastic to control melting and using natural currents to assist in the transportation process, so the idea to harvest water from icebergs continues to have supporters.

Atmospheric vapor as a water source

The most direct way to collect atmospheric water vapor is to harvest precipitation before it reaches the ground

Figure 7. Icebergs are routinely towed to avoid collisions with oil drilling platforms and ships. Although currently impractical, recent research shows that icebergs could someday provide freshwater for dry coastal areas.

and is lost by infiltration into the ground or by runoff into streams, lakes, or oceans. A common approach is to catch precipitation flowing from rooftops and store it in ponds or tanks for later use. This is an effective way to supplement existing water supplies, but does not provide reliable quantities and cannot be used for drinking without expensive filtering systems.

In areas with little or no precipitation, it may be possible to extract usable quantities of water vapor directly from the atmosphere. This is often done by mimicking processes used by other organisms to extract atmospheric water vapor. For example, tropical rainforest plants called **epiphytes** that live attached to the branches of trees far above the forest floor use specialized scales on their leaves to capture and absorb water from the humid air.

Figure 8. Resembling huge volleyball nets, these fog collectors provided freshwater for hundreds of residents of the coastal fishing village of Chungungo, Chile.

Humans can also capture atmospheric water using simple devices called fog-water collection systems. These devices are basically large, fine nets that catch and collect the small airborne droplets of water that make up fog. Figure 8 shows one of these systems currently in use in Chile's Atacama Desert that produces about 9500 liters of water per day. This system doubled the amount of water available to the community and eliminated the need for hauling water in by truck.

A scientist in Germany recently developed a system that uses new materials that "grab" airborne water, greatly increasing the amount of water vapor caught for a given surface area. Fog-water collection represents a promising solution for obtaining water in small, isolated communities near large bodies of saltwater.

6. What technologies might be needed to make it practical to use the ice sheets or icebergs as sources of freshwater? Explain.

7. How might differences in climate (dry or humid, hot or cold) affect the possibility of harvesting water from the atmosphere?

Global warming

Global warming—in which the Earth's mean temperature rises due to many different factors over geological time—is an important concern today. The phenomenon is a complex issue, both scientifically and politically, and there have been numerous studies and counter-studies that support and refute the claim that global warming is occurring. One consequence of global warming is increased melting of the polar ice sheets. In the next investigation you will examine the impact of such an event.

8. If the ice sheets were to melt, how might that affect the other water reservoirs? Explain.

Investigation 1.4

What if the ice sheets melted?

Increasing global temperatures, or global warming, is an important concern today. In this activity, you will examine the potential consequences of an extreme global warming event in which the Antarctic and Greenland ice sheets melt.

1. a. What do you think would happen to the oceans and continents if the ice sheets melt?

Sea level and melting ice sheets

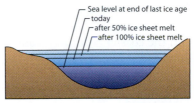

Changes in sea level caused by changes in global ice sheets (not to scale).

b. How would these changes affect our society?

The change in mean sea level

To estimate the change in sea level that would occur if the ice sheets melted, you will determine the volume of the ice sheets and then calculate how that water would change the depth of the oceans.

2. In Investigation 1.2, you determined the total volume of water in the ice sheets (Table 2, page 10) and the total surface area of the oceans (Table 1, page 9). Write those values below, including the proper units of measurement.

a.

What values?

If you did not do Investigation 1.2, your instructor will give you these values.

b.

If the ice sheets melted, most of the water would enter the oceans, raising sea level. As water flooded low-lying coastal regions, the oceans' surface area would increase. To simplify the calculation in the sea-level change, however, assume this change in the surface area would be small enough to ignore.

Calculating a change in sea level

Change in sea level (km) =

$$\frac{\text{Ice sheet water volume (km}^3)}{\text{Surface area of oceans (km}^2)}$$

3. a. Calculate the change in sea level (in km) if the ice sheets melted, and write your result below (see sidebar).

b. Convert your result from kilometers to meters by multiplying the result in part 3a by 1000.

Effects of rising sea level

Now that you have a rough estimate of the change in sea level due to the complete melting of the ice sheets, you can examine the effects of a sea-level rise on global land masses.

Fastest glacier?

If you would like to learn more about how Greenland's ice sheet is responding to global warming, you can view a video clip from **NOVA Science Now** on their Web site, at

**www.pbs.org/wgbh/nova/
sciencenow/3210/03.html**

Changing coastlines

💻 Launch ArcMap, then locate and open the **ewr_unit_1.mxd** file.

Refer to the tear-out Quick Reference Sheet located in the Introduction to this module for GIS definitions and instructions on how to perform tasks.

💻 In the Table of Contents, right-click the **Melting Ice Sheets** data frame and choose Activate.

💻 Expand the **Melting Ice Sheets** data frame.

This data frame shows the countries of the world and the coastal areas that would be flooded by a 10-m, 30-m, and 60-m rise in sea level.

4. On Map 1 below, circle two major regions that would experience significant flooding from a 60-m rise in sea level.

Map 1 — Regions predicted to experience significant flooding

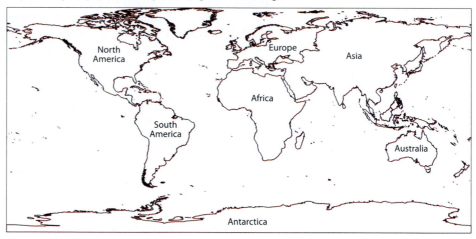

💻 Use the Zoom In tool 🔍 to zoom in on the United States (excluding Alaska and Hawaii).

5. Near which U.S. coast (east or west) would you rather be living if the ice sheets were to melt? Why?

Identifying states

If you need help identifying states, make sure the **U.S. States** layer is on. Using the Identify tool ⓘ, click on a state to read its name. (Be sure to select **U.S. States** from the drop-down list of layers in the Identify Results window.)

💻 Turn on the **U.S. States** layer.

6. After a 60-m sea-level rise, which states are mostly under water? (If you need help identifying states, see note at left. Use the Zoom In tool 🔍 as necessary).

💻 Close the Identify Results window if necessary.

🖥 Click the Full Extent button 🌐 to view the entire map.

🖥 Turn off the **10-meter Rise**, **30-meter Rise**, and **60-meter Rise** layers.

🖥 Turn on the **Affected Population** layer.

Impact on coastal cities

The **Affected Population** layer shows the global population living within the region affected by a 60-m increase in sea level. The layer consists of small rectangles, each of which is color-coded according to its population. Areas with fewer than 10,000 inhabitants are not shown.

🖥 Use the Zoom In 🔍 and Pan ✋ tools to examine the population data more closely.

7. In which countries will the most people be affected by a 60-m rise in sea level?

Identifying countries

If you need help identifying countries, make sure the **Countries** layer is on. Using the Identify tool 🛈, click on a country to read its name. (Be sure to select **Countries** from the list of layers in the Identify Results window.)

8. Are these countries in the same geographic regions that you identified in question 4? Explain.

🖥 Click the Full Extent button 🌐 to view the entire map.

Next, you will determine the total population affected by a sea-level rise of 10, 30, and 60 meters, as well as the number of cities and the amount of cropland that would be flooded.

Estimating flooded area

Determine the total area flooded for each rise in sea level, starting with the **10-meter Rise** layer. This layer shows the land area that would be flooded by a 10-m rise in sea level.

🖥 Turn on the **10-meter Rise** layer.

🖥 Click the Statistics button 🗙.

🖥 In the Statistics window, calculate statistics for **all features** of the **10-meter Rise** layer, using the **Area (km^2)** field.

🖥 Click **OK**. Be patient while the statistics are calculated.

The total area flooded is reported in the Statistics window as the **Total**.

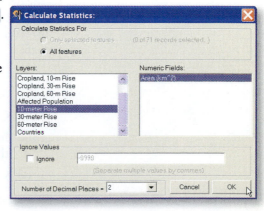

9. Record the total area (**Total**) in the **Total flooded area (km²)** column of Table 1. Round to the nearest 100,000 km².

Table 1 — Impacts if sea level rose 10, 30, or 60 meters

Sea-level rise	Total flooded area km² to nearest 100,000	Total population affected to nearest 100,000,000	Number of cities flooded	Total cropland flooded km² to nearest 100,000
10 meters				
30 meters				
60 meters				

🖥 Close the Statistics window.

🖥 Repeat the above steps to find the total area flooded by 30- and 60-m increases in sea level. Round your results to the nearest 100,000 km² and record them in Table 1.

🖥 Close the Statistics window when you are finished.

Estimating the affected population

Next, estimate the total population that would be affected in each of the flooded areas.

🖥 Click the Select By Location button 🖳.

🖥 In the Select By Location window, construct the query statement:

I want to **select features from** the **Affected Population** layer that **intersect** the features in the **10-meter Rise** layer (*not the Cropland, 10-m Rise layer*).

🖥 Click **Apply**.

🖥 Close the Select By Location window.

The densely populated areas that would be flooded by a 10-m rise in sea level should be highlighted on the map.

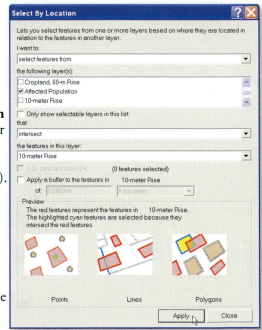

🖥 Select the **Affected Population** layer.

🖥 Click the Statistics button Ⓧ.

🖥 In the Statistics window, calculate statistics for **only selected features** of the **Affected Population** layer, using the **Population** field.

🖥 Click **OK**. Be patient while the statistics are calculated.

The total number of people affected by a 10-m rise in sea level is reported in the Statistics window as the **Total**.

10. Round the total population affected by a 10-m sea-level rise to the nearest 100,000,000 and record it in Table 1 on the previous page.

🖥 Close the Statistics window.

🖥 Repeat the Select By Location and Statistics procedures to determine the total population that would be affected by 30- and 60-m increases in sea level and record the totals in Table 1 on the previous page.

🖥 Close the Statistics window when you are finished.

🖥 Click the Clear Selected Features button ⊠.

🖥 Turn off the **Affected Population** layer.

Flooded cities

Next, you will determine the number of cities affected by flooding due to a rise in sea level.

🖥 Turn on the **Major Cities** layer.

🖥 Select the **Major Cities** layer.

The **Major Cities** layer includes cities with populations greater than 20,000, capitals, and small but geographically important towns. The size of the circle represents the size of the city's population. Determining the total number of cities affected by a sea-level rise is similar to estimating the populations affected by a sea-level rise.

🖥 Click the Select By Location button 🔲.

🖥 In the Select By Location window, construct the query statement:

I want to **select features from** the **Major Cities** layer that **intersect** the features in the **10-meter Rise** layer.

🖥 Click **Apply**.

🖥 Close the Select By Location window.

The number of major cities that would be flooded by a 10-m rise in sea level should be highlighted on the map. To quickly determine the number of flooded cities:

🖥 Click the Open Attribute Table button 🔲.

🖥 Read the total number of major cities that would be flooded at the bottom of the table (your answer will be different than the example shown below).

read number here

Show: | All | Selected | Records (0 out of 19 Selected.)

11. Record the number of cities in Table 1 on the previous page.

🖥 Close the attribute table.

🖥 Repeat the Select By Location procedure to determine the number of cities flooded by 30- and 60-m rises in sea level and record the results in Table 1.

🖥 Close the Select By Location window and the attribute table when you are finished.

🖥 Click the Clear Selected Features button ⊠.

12. If sea level rose 10 m, would a major city within your state be flooded? (Use the Zoom In tool 🔍 if you need to see your state in more detail.) Even if you do not live in a city that would be flooded, how could flooding in other major cities affect you?

💻 Turn off the **Major Cities** layer.

💻 Turn off the **10-meter Rise**, **30-meter Rise**, and **60-meter Rise** layers.

Flooded croplands

Finally, you will determine the amount of cropland affected by flooding, using a series of layers that show the intersections between areas that would flood and the world's croplands.

💻 Turn on the **Cropland, 10-m Rise** layer.

This layer shows the area of cropland that would be flooded due to a 10-m rise in sea level.

💻 Click the Statistics button 🗵.

💻 In the Statistics window, calculate statistics for **all features** of the **Cropland, 10-m Rise** layer, using the **Area (km^2)** field.

💻 Click **OK**. Be patient while the statistics are calculated.

The total area of cropland flooded by a 10-m rise in sea level is given as the **Total**.

13. Record the total area of cropland flooded in Table 1 on page 22. Round to the nearest 100,000 km².

💻 Close the Statistics window.

💻 Repeat the Statistics procedure, using the **Cropland, 30-m Rise** and the **Cropland, 60-m Rise** layers, to find the total area of cropland flooded by 30- and 60-m rises in sea level.

💻 Close the Statistics window when you are finished.

By themselves, the numbers you calculated on land area, population, and cropland affected by a rise in sea level are hard to assess. It is useful to compare these numbers to the total land area, population, and amount of cropland in the world. You can do this by calculating the *percentages* of each of these categories affected by a rising sea level. These totals are given in Table 2.

Table 2 — Global statistics

Total land area	147,000,000 km²
Total population	6,200,000,000 people
Total major cities	606
Total cropland area	23,000,000 km²

What if the ice sheets melted?

Calculating percentages

To calculate percentages for Table 3, divide the value from the corresponding cell in Table 1 by the appropriate total in Table 2, and multiply by 100. For example,

$$\% \text{ Area flooded} = \frac{\text{Total flooded area (km}^2)}{\text{Total land area (km}^2)} \times 100$$

14. Use the totals provided in Table 2 to calculate the percentage affected for each category for a 10-, 30-, and 60-m rise in sea level and complete Table 3. Round all results to the nearest 1 percent. (See sidebar for an example of how to calculate a percentage).

Table 3 — Percent area, population, and cropland affected by sea-level rise

Sea-level rise	% Area flooded	% Population affected	% Major cities flooded	% Cropland flooded
10 meters				
30 meters				
60 meters				

Plotting your results on graphs will help you interpret them. First, you will examine the relationship between the percentage of affected land area and the percentage of affected population.

15. On Graph 1, plot your results for the percent of population affected (on the vertical axis) versus your results for the percent of land area flooded (on the horizontal axis). Connect your points with a smooth solid curve that passes through the origin of the graph (0% land affected, 0% of the population affected).

Graph 1 — Effects of sea-level rise on population and cropland

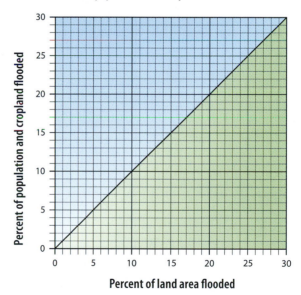

(vertical axis) Percent of population and cropland flooded

(horizontal axis) Percent of land area flooded

16. Plot the percent of cropland flooded versus the percent of total land area flooded on the same graph, and connect your points with a smooth dashed curve that begins at the origin of the graph.

The diagonal line divides the graph into two regions that show the relative effect of a rising sea level on land area, population, and cropland. Points plotted above the line (in the blue region) show a greater impact of a rising sea level on people and cropland. Points plotted below the line (in the green region) show a greater impact of a rising sea level on total land area.

17. Which is affected more strongly by a rising sea level, population or land area? What does this tell you about where people tend to live—near coastlines or in the interiors of continents? Explain.

Rising water and global population

The preceding scenarios in this investigation are extreme. A global rise in sea level of one to two meters is a more realistic hazard. You can use your data to construct a graph that will help estimate the effects of such a scenario.

18. On Graph 2, plot the percent of the world's population affected versus the sea-level rise. Use the numbers in Table 3 on the previous page to plot these points. When finished, you will have four points on the graph, each associated with a 0-, 10-, 30-, or 60-m sea-level rise. Connect your points with a smooth curve that passes through the origin of the graph.

Graph 2 — Percent of population affected versus sea-level rise

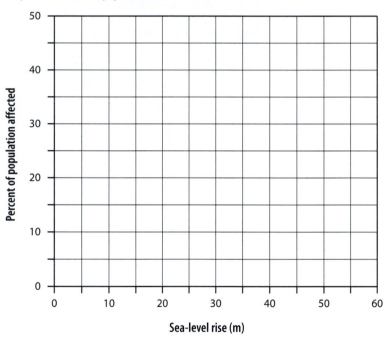

19. According to Graph 2, what percent of the world's population would be affected by a 2-m rise in sea level? How about a 5-m rise?

20. Are these flood scenarios a cause for concern for the majority of the world's population? Explain your answer.

🖥 Quit ArcMap and do not save changes.

Assessing the accuracy of your calculation

At the beginning of this activity, you estimated the rise in sea level due to the melting of the continental ice sheets. One of the assumptions you made was that the change in the surface area of the ocean was small enough to ignore. Later, however, you saw that the amount of land covered by the rising sea was quite substantial. In fact, in the 60-m rise scenario, the amount of land covered by floodwaters is about half the size of the Atlantic Ocean. You can use this knowledge to refine your estimate of the rise in sea level.

21. What is the combined area of the oceans and the 60-m floodwaters? (Refer to question 2b and Table 1 in this investigation. Be sure to include your units).

22. Calculate the rise in sea level that would occur if the continental ice sheets melted completely.

   a. Divide the total volume of water in the ice sheets (question 2a in this investigation) by the combined area (question 21).

   b. Multiply the above result by 1000 to convert from kilometers to meters, and round to the nearest meter.

23. You now have two estimates for the amount of sea-level rise if the ice sheets melted: your original estimate (question 3b in this investigation), and the estimate you calculated above, which accounts for a change in the surface area of the oceans.

   a. Which of these estimates represents the *upper limit*, or higher estimate, of sea-level rise?

   b. Which of these numbers represents the *lower limit*, or lower estimate, of sea-level rise?

   c. What is a more likely estimate for sea-level rise, given these upper and lower limits?

Comparing major reservoirs

In this unit, you have examined global water reservoirs, including lakes, rivers, groundwater, the atmosphere, seas and oceans, and ice sheets. Nearly all of the freshwater used around the world currently comes from surface and groundwater reservoirs. Water from these sources has been plentiful and relatively easy and inexpensive to obtain.

Water in the atmosphere

When you consider the atmosphere as a water reservoir, be sure to include water vapor, suspended droplets, and precipitation.

As the global population grows, the demands on the world's surface and groundwater resources are increasing. In many areas the quality of water is declining as pollution contaminates the water supply with toxic chemicals and waterborne diseases. Providing an adequate supply of safe, usable water is becoming increasingly difficult and expensive. In the future, we will need to improve efforts to keep water supplies safe and clean and to utilize other water reservoirs to meet global demand.

1. Form a team and brainstorm the advantages and disadvantages of using each of the global water reservoirs listed in the top row of Table 1 on page 31. The criteria listed in the first column of the table are a guide for issues that you may need to consider. Try to consider all of the possible consequences and benefits of using each water source. Be prepared to share your team's ideas with your class.

2. Research the practicality of using the atmosphere, oceans, and ice sheets as sources of freshwater. Are any of these reservoirs currently in use? If so, by whom? If not, why do you think they are not used?

   a. Atmosphere

   b. Oceans

   c. Ice sheets

3. Describe the characteristics of a country that would be most likely to utilize the global water reservoirs listed below as sources of freshwater.

   a. Atmosphere

   b. Oceans

   c. Ice sheets

The World Factbook

The World Factbook can be found at the CIA Web site at:

http://www.cia.gov/cia/publications/factbook/

4. List the name of one country that you think best meets the characteristics you listed in question 3 for utilizing each reservoir. Briefly explain why you think that reservoir is that country's best option for supplying its freshwater needs. (You may use the data in the **ewr_unit_1** ArcMap project file along with an atlas, the World Factbook, or other reference sources to answer this question).

   a. Atmosphere

   b. Oceans

   c. Ice sheets

Table 1 — Comparison of global water reservoirs

Usability criteria	Oceans	Atmosphere	Ice sheets
Accessibility and delivery			
Quality			
Cost			
Sustainability			
Environmental impact			

Unit 2
The Renewable Resource

In this unit, you will

- *Discover the importance of precipitation to our water supply.*

- *Examine global and regional precipitation patterns.*

- *Investigate the factors that determine where precipitation falls and how it moves when it reaches land.*

Bonneville Power Administration

Bonneville Dam spans the Columbia River 65 kilometers upstream from Portland, Oregon.

Too little, too much

Weather — the state of the atmosphere at a given time and place, including temperature, moisture, wind, and pressure.

Climate — weather patterns over a long period, or "average" weather for a particular place. For example, the climate of the southwestern U.S. is generally warm and dry compared to other parts of the country.

"Climate is what we expect...and weather is what we get."

Robert Heinlein

How common is drought?

At any given time, an average of about 21 percent of the contiguous U.S. is experiencing moderate to extreme drought.

In the previous unit you examined global water reservoirs and estimated the volume of water each contains. In this unit you will investigate the ways in which water moves between these global reservoirs, focusing on the two most visible paths of the hydrologic cycle, precipitation and runoff. Precipitation replenishes surface, soil, and groundwater resources critical to human society, while runoff in streams and rivers provides power and transportation and distributes water resources from regions with abundant water to regions where water is scarce.

Precipitation and runoff are influenced by weather and climate patterns. Because both precipitation and runoff can be extremely variable when examined over the short term (days to months), this unit will focus on investigating long-term (30-year) precipitation and runoff patterns. You will identify factors that influence when and where precipitation occurs in the United States, and what happens to that precipitation after it reaches Earth's surface. Short-term precipitation and runoff events such as storms and floods can have dramatic economic and societal consequences, so you will begin this unit by considering major U.S. weather disasters.

Too little water

The Great Depression started with the crash of the stock market in 1929. By 1933, 25 – 30 percent of the workforce was unemployed and the stock market had lost 80 percent of its value. To recover from economic losses suffered during the peak years of the Depression, farmers plowed more land, planted more crops, and grazed their pastures more intensively, causing extensive soil damage. This damage, combined with a prolonged drought, caused a series of dust storms or *black blizzards* to sweep across the central plains. The winds removed millions of tons of topsoil from barren fields and darkened skies across the country. Farmers went bankrupt, and vast stretches of farmland were abandoned. The region hit hardest during this period was given the name *the Dust Bowl*.

To appreciate the magnitude of this event, read the eyewitness accounts on pages 37 – 42. The first story describes the dust storm of April 14, 1935, known as *Black Sunday*. The second account is an excerpt from the diary of Ann Marie Low, describing life on a North Dakota farm during the Dust Bowl years.

The Dust Bowl was a significant event, but not unique. Major droughts are common in the Midwest and in other parts of our nation and the world. The story on pages 42 – 44 describes the effects of a recent drought that struck Kansas during the summer of 2002.

Too much water

When rain falls or snowpack melts faster than the surface can absorb water, the excess water flows downhill over the surface as runoff. If the amount or extent of runoff is unusually high, flooding may occur. Rather than affecting the areas where the precipitation fell, floods often occur downstream from areas that experience heavy precipitation. The story on pages 44–47 tells about a recent major flood event, the Mississippi River floods of 1993.

Billion-dollar disasters

Most natural disasters that strike the United States are the result of having too little or too much water. The map below shows weather-related disasters in the U.S. between 1980 and 2005 that resulted in damage costing a billion dollars or more.

    1. Study the map and its legend below. What types of natural disasters are associated with too much water, and where do they generally occur in the U.S.?

    2. What types of natural disasters are associated with a lack of precipitation, and where do they generally occur in the U.S.?

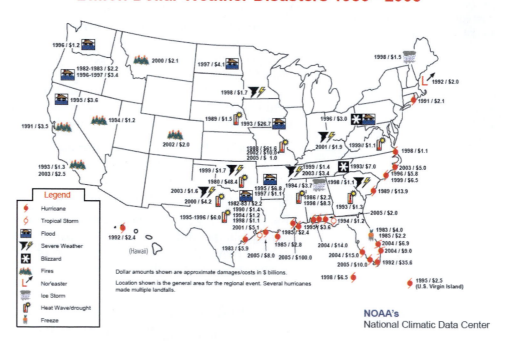

Figure 1. Billion dollar weather disasters, 1980–2005.

3. On the map, locate the four most costly weather disasters. Record the year, region (use map in sidebar to determine region), type of disaster, and cost in Table 1 below.

Regions of the U.S.

Table 1 — Most costly U.S. weather disasters, 1980–2004

Year	Region (states)	Type	Cost ($ billions)

4. Even if these disasters did not occur near you, how could they have affected you or your family?

Eyewitness accounts

Eyewitness accounts of historical events are valuable research tools. They not only help to establish and clarify facts, but also serve to put the event in a human context. The following stories, taken from books and newspapers, provide a glimpse of the human impact of droughts and floods.

The Black Sunday dust storm (1935)

'Worst' Duster Whips Across Panhandle
Norther Strikes Sunday to Blot out Sun, Turn Day into Night

AMARILLO, Texas, April 15 (AP) — North winds whipped dust of the drought area to a new fury Sunday and old timers said the storm was the worst they'd seen. A black duster — sun-blotting cloud banks — raced over Southwest Kansas, the Texas and Oklahoma Panhandles, and foggy haze spread about other parts of the southwest.

A gentle, north breeze preceded 8,000-feet-high clouds of dust. As the midnight fog arrived, the streets were practically deserted. However, hundreds of people stood before their homes to watch the magnificent sight.

Darkness settled swiftly after the city had been enveloped in the stinking, stinging dust, carried by a 50-mile-an-hour wind. Despite closed windows and doors, the silt crept into buildings to deposit a dingy, gray film. Within two hours the dust was a quarter of an inch in thickness in homes and stores.

A wall of dust approaches a Kansas town in 1935. Dust clouds often reached speeds of up to 60 miles per hour (95 kilometers per hour) and were a mile (1.5 kilometers) or more high.

The storm struck just before early twilight. All traffic was blocked and taxi companies reported that it was difficult to make calls for nearly 45 minutes. Street signal lights were invisible a few paces away. Lights in 10- and 12-story buildings could not be seen.

John L. McCarty, editor of the *Dalhart Texan*, of Dalhart, the center of the drought-stricken area of the Panhandle, called a few minutes before the storm arrived in Amarillo. "I went outside the house during the storm and could not see a lighted window of the house three feet away." Mr. McCarty said.

Damage to the wheat crop, already half ruined by drought and wind, could not be learned last night, but several grain men believed that the dust would cover even more of the crops.

Left-hand photo was taken in 1935 in Garden City, Kansas at 5:15 p.m. Right-hand photo, taken 15 minutes later, shows the city during a dust storm that turned day into night, blotting out the sun. Note the street lights for orientation of the two photos.

5. In the 1930s, how did people learn that a large dust storm was coming toward them? How would they get this news today?

Dust Bowl diary

The following excerpts are from the diary of Ann Marie Low, whose family lived on a farm in North Dakota during the Dust Bowl years.

Reprinted from *Dust Bowl Diary* by Ann Marie Low with permission of the University of Nebraska Press.

Copyright © 1984 by the University of Nebraska Press.

April 25, 1934, Wednesday

Last weekend was the worst dust storm we ever had. We've been having quite a bit of blowing dirt every year since the drouth [drought] started, not only here, but all over the Great Plains. Many days this spring the air is just full of dirt coming, literally, for hundreds of miles. It sifts into everything. After we wash the dishes and put them away, so much dust sifts into the cupboards we must wash them again before the next meal. Clothes in the closets are covered with dust.

Last weekend no one was taking an automobile out for fear of ruining the motor. I rode Roany [Ann's horse] to Frank's place to return a gear. To find my way I had to ride right beside the fence, scarcely able to see from one fence post to the next.

Newspapers say the deaths of many babies and old people are attributed to breathing in so much dirt.

May 21, 1934, Monday

Saturday Dad, Bud, and I planted an acre of potatoes. There was so much dirt in the air I couldn't see Bud only a few feet in front of me. Even the air in the house was just a haze. In the evening the wind died down, and Cap came to take me to the movie. We joked about how hard it is to get cleaned up enough to go anywhere.

George E. Marsh Album, NOAA

Dust cloud approaches Stratford, Texas on April 18, 1935.

The newspapers report that on May 10 there was such a strong wind the experts in Chicago estimated 12,000,000 tons of Plains soil was dumped on that city. By the next day the sun was obscured in Washington, D.C., and ships 300 miles out at sea reported dust settling on their decks.

May 30, 1934, Wednesday

The mess was incredible! Dirt had blown into the house all week and lay inches deep on everything. Every towel and curtain was just black. There wasn't a clean dish or cooking utensil. There was no food.

It took until 10 o'clock to wash all the dirty dishes. That's not wiping them — just washing them. The cupboards had to be washed out to have a clean place to put them.

Mama couldn't make bread until I carried water to wash the bread mixer. I couldn't churn until the churn was washed and scalded. We just couldn't do anything until something was washed first. Every room had to have dirt almost shoveled out of it before we could wash floors and furniture.

We had no time to wash clothes, but it was necessary. I had to wash out the boiler, wash tubs, and the washing machine before we could use them. Then every towel, curtain, piece of bedding, and garment had to be taken outdoors to have as much dust as possible shaken out before washing. The cistern is dry, so I had to carry all the water we needed from the well.

That evening Cap came to take me to the movie, as usual. I'm sorry I snapped at Cap. It isn't his fault, or anyone's fault, but I was tired and cross. Life in what the newspapers call "the Dust Bowl" is becoming a gritty nightmare.

August 1, 1934, Wednesday

The drouth and dust storms are something fierce. As far as one can see are brown pastures and fields which, in the wind, just rise up and fill the air with dirt. It tortures animals and humans, makes housekeeping an everlasting drudgery, and ruins machinery.

The crops are long since ruined. In the spring wheat section of the U.S., a crop of 12 million bushels is expected instead of the usual 170 million. We have had such drouth for five years all subsoil moisture is gone. Fifteen feet down the ground is dry as dust. Trees are dying by the thousands. Cattle and horses are dying, some from starvation and some from dirt they eat on the grass.

October 1, 1934, Monday

Woodger, the federal acquisition agent, finally got around to see Dad last Saturday. From the kitchen I could hear the whole conversation, and it amused me. Dad was sitting on the back steps, resting after the noon meal, when a government car drove up. In this country, when anyone drives in, you meet him at the gate with hand outstretched, making him welcome. Dad knew this was the agent who had been dealing with the banks to rob our neighbors of their land. He didn't get up.

Dust storms often arose out of nowhere on otherwise beautiful days. Here, a car seems to be trying to outrun an approaching dust storm.

Arthur Rothstein, FSA-OWI Collection, Library of Congress

Woodger, a small man with a toothbrush-shaped mustache, walked up to the steps and introduced himself. He told about the proposed refuge and said he was there to appraise Dad's land and make on offer for it.

"It is not for sale."

In a sneering and condescending tone, "Oh, I believe this whole area is. All your neighbors are selling."

"No, they aren't. The banks are selling their places from under them. This place is not mortgaged."

Woodger spoke of what a great benefit this wildlife refuge will be — something for the good of all the people.

Sand and dust buried farms and equipment, killed livestock, and caused human suffering during the Dust Bowl years.

"I didn't build this ranch up for the benefit of all the people, but for me and my family."

Woodger pulled out every argument he could think of, then finally said, "After all, this is submarginal land on which you can't make a living."

That was news to Dad. He stood up, very slowly. The little pipsqueak agent stared in amazement as the bulk of Dad loomed above him. Dad is six feet four inches in his stocking feet, and higher with his boots on. He weighs 250 pounds, mostly bone and muscle. He is so big-boned and broad-shouldered it takes 250 pounds to flesh him out properly. Because his face has kept firm flesh and his hair has stayed jet black, he looks far younger than he is. His blue eyes, startling in his swarthy face, have never lost their keenness. By the time he had drawn himself up to his full height, the agent was open-mouthed.

"Young man, I want to tell you something. I've been here since the Territorial days. I started out with the clothes on my back and a $10.00 gold piece. I was young and dumb and uneducated. I didn't know I couldn't make a living here, and I didn't have any government expert to tell me so.

"Young man, I've been fighting drouth and depression and blizzard and blackleg ever since the Territorial days. Everything you can see from here to the horizon belongs to me—the land, cattle, buildings, horses, and machinery. It is too late for you to tell me I can't make a living here. You better go away before you make me mad."

As Woodger scuttled for his car, Dad called after him, "By the way, when you get back to Washington, D.C., you can tell Franklin Delano Roosevelt I still have that $10.00 gold piece, too!"

Having gold is illegal now.

If Dad can get a decent price, he probably ought to sell. He is getting too old for a spread like this. Bud doesn't want it. Mama doesn't like it here. I love it, but am not going to. Everything I loved will be gone.

[Two years later…]

August 1, 1936, Saturday

July has gone, and still no rain. This is the worst summer yet. The fields are nothing but grasshoppers and dried-up Russian thistle [tumbleweeds]. The hills are burned to nothing but rocks and dry ground. The meadows have no grass except in former slough holes, and that has to be raked and stacked as soon as cut, or it blows away in these hot winds. There is one dust storm after another. It is the most disheartening situation I have seen yet. Livestock and humans are really suffering. I don't know how we keep going.

The dirt quit blowing today, so I cleaned the house. What a mess! The same old business of scrubbing floors in all nine rooms, washing all the woodwork and windows, washing the bedding, curtains, and towels, taking all the rugs and sofa pillows out to beat the dust out of them, cleaning closets and cupboards, dusting all the books and furniture, washing the mirrors and every dish and cooking utensil. Cleaning up after dust storms has gone on year after year now. I'm getting awfully tired of it. The dust will probably blow again tomorrow.

6. How was Ann's daily life affected by the drought? Give specific examples from her diary.

The Drought of June 2002

Kansas drought devastates wheat crop, forces widespread liquidation of cattle

By Roxana Hegeman

ELKHART, Kansas (AP)—Warren Bowker's combine kicks up a cloud of dust as he runs it nearly full speed across his thin stands of winter wheat. The machine almost touches the parched ground as it tries to cut stunted wheat that grew only a few inches tall. Bowker's brother, Shaun, waves him in. Moments later they stare glumly at the combine's flat tire. Shaun Bowker uses his cell phone to call a repair shop, which says someone will be out soon. After all, there isn't much business these days.

It's been nearly a year since much of western Kansas has gotten substantial rain of even up to an inch, and the southwest corner has been hardest hit. The drought has devastated the wheat crop now being harvested and spurred widespread selling off of cattle herds, as farmers become increasingly desperate to find enough feed and water to carry them through the summer grazing season. Rural farm economies are hurting and even the wildlife is struggling to survive. The Bowker brothers are thankful to have anything left at all to harvest.

Arthur Rothstein, FSA-OWI Collection, Library of Congress

A farmer and his sons walk through a dust storm in Cimarron County, Oklahoma in April 1936.

Poor crops are the least of their worries in this drought. Before the end of the month, the Bowkers will round up their cattle out of the Cimarron National Grassland and ship them off for sale, liquidating in one day what took them 10 years to build. "We are going to dump the whole thing. We aren't going to fight it," Warren Bowker said.

Last week, forestry officials ordered all 100 farmers with permits to graze government lands to remove their grazing cattle from the drought-stressed grass. Usually 5,000 cattle feed off the national grassland; 3,200 are on it now, and all must leave before the end of June. "The grass and vegetation is stressed so severely that to graze it will be detrimental," says Cimarron National Grassland district manager Joe Hartman.

Weather records dating back to 1913 show that never has there been less precipitation here than now. Even the Dust Bowl days of the 1930s logged more rain than this year, says Morton County Extension agent Tim Jones. The big, black dust clouds of that era haven't repeated because much of the land has been put into the Conservation Reserve Program, a government program that pays farmers not to plant their cropland. But at times, big drifts of dirt blow across state highways so thickly that for a moment it seems like dusk. The drifting soil piles up along fence rows.

Activity at the Elkhart Co-op grain elevator — or lack of it — illustrates the troubles. Manager Larry Dunn says his seven elevators usually take in 3.2 million bushels of wheat during harvest. This year they hope to collect 500,000 to

600,000 bushels. He figures 70 percent of the planted acres were abandoned long before harvest began. "It is to the point it can get easily depressing for employees who have to hear it all the time," Dunn says.

Roughly 2,800 Kansas farmers have filed insurance claims for this year's crop, collecting $24 million so far even as losses mount with the start of the harvest, according to figures compiled by the federal Department of Agriculture's risk management agency. Those figures only reflect claims paid, and the agency has a backlog. They don't include damage from a recent weekend hail storm that caused an estimated $6 million in damages to wheat crops. This year's wheat crop in Kansas is insured for $645 million, and the money paid out so far is mostly for abandoned acres, says Rebecca Davis, the agency's director of the Topeka regional office.

"We are all trying to stay optimistic, but it is kind of bleak," says Pam Pate of Ben Pate Agency in Elkhart, noting that about 75 percent of the farmers who bought insurance from the agency have already filed claims for abandoned acres. "It will turn around and get good again," she says. "We are hoping prices will come back up. It will rain or snow again. We are tough out here. We survive."

Elkhart has been through droughts before. Businesses come and go, but it will be mainly farmers who are forced to quit. "A lot of our customers have no wheat left to cut," says Tim Predmore, service manager at a John Deere farm equipment dealership. "As far as we are concerned, there is no harvest."

7. Aside from farming, what other types of businesses can be affected by drought?

The 1993 Mississippi Flood

In 1993, only three years after a severe drought caused billions of dollars in agricultural and industrial damage, a major flood occurred on the Upper Mississippi and Lower Missouri Rivers. So widespread were the affects of this flood that it has come to be called "The Great Flood."

Small Town Paid Its Dues Like Others Along The River

By Mike Harden; Dispatch Columnist

NUTWOOD, Illinois, July 21, 1993 (The Columbus Dispatch) — With his boots firmly planted on the half-submerged road outside Joyce's Kountry Korner, 1st. Lt. Steve Carney had the wary look of a man half-expecting a sucker punch.

But another wallop from the brown swirl of debris-laden flood water was hardly necessary. By yesterday, there was little left to do but tally the losses and clean up the mess.

Most of the nation was contemplating St. Louis' battle to save its south side from the surge of the Mississippi when little Nutwood went under. Just up the mouth of the Illinois River from the Gateway Arch, the tree-nestled village of 100 or so, wasn't much of a magnet for the TV crews and their satellite dish-topped trucks. But its fight was no less gallant than that in St. Louis.

An aerial view shows the extent of flood damage from the 1993 flood.

Sandbags are filled by residents and volunteers to stop further damage by rising flood waters during the 1993 flood.

Side-by-side with townsfolk, Illinois National Guard company commander Carney and the troops of Bravo Company, 133rd Signal Battalion, threw everything they had at the gorged Illinois River.

As late as Sunday afternoon, it appeared they might save Nutwood. The river, after all, would have to breach two levees and cut its way through 10,000 acres of prime farmland in the Nutwood Drainage and Levee District to even touch the tiny hamlet.

Carney recalls the instant he knew it had done just that. With water lapping over the two main levees early Sunday, he ordered his troops to climb on the bulldozers and push a dirt parapet up around the sandbagged outskirts of Nutwood to provide yet one more obstacle in case the worst happened.

It had, he knew, when he peered through his binoculars toward the main levee and saw two figures running up the road toward the town. "There was a gap 150 yards long in the levee," he said.

Moving like a surge of lava, the water was chasing the pair toward Nutwood. Carney jumped in his vehicle to rescue them, but a National Guard helicopter swooped in and plucked them to safety.

"Down the left-hand side of the road," Carney recalled, "it looked like the Niagara."

"It sounded like a train."

It was clear that the hastily erected parapet would only slow the inevitable.

"We were trying to buy time for the people to get their stuff out and evacuate," he continued. "We told them, 'Get your pictures and your family Bibles and enough clothes to stay a couple days.'"

Miles away, at the Jersey County Fair, a call went out over the public-address system, appealing for any available vehicles to aid in moving household possessions out of the stricken hamlet.

Homes, businesses, and personal property were all destroyed by the high flood levels during the 1993 flood. These homes survived with little damage — others were not so fortunate.

Too little, too much   45

An eerie calm settles on the water in this neighborhood in East Grand Forks, Minnesota. Many homes floated off their foundations, and all were significantly damaged.

Off in the distance, up the dusty, sinuous road that winds down into Nutwood from nearby Fieldon, a long string of pickup trucks—cavalry reinforcements—poured in from the county fair to help.

The water surged through the corn rows and into houses and mobile homes.

With all the people out and safe, the National Guard took care of one last, small lifesaving job: It unleashed two dogs, forgotten in the panic and still chained in yards.

Just around the bend, on a farm outside town, 78-year-old Harold Klunk made a last-ditch effort to rescue more than 3,000 bushels of corn that were stored in bins directly in the water's path.

He was aided in the frantic task by his nephew, farmer Jim Vahle, and Jim's sons, Matt and Gary.

With one of the last of the grain trucks loaded as the Monday sun was dipping toward dusk, a spry Klunk pulled himself up into the driver's seat and checked the rear view mirror.

Behind him lay 10,000 acres of submerged crops—his and all of his neighbors'. In the 75 years that had passed since he had been brought to the farm that had once been his father's, he had seen nothing like it.

In the yard, Vahle's wife, Wanda, fetched cold drinks for her husband and sons and watched, mildly amused, as mice and small rats scampered across floating twigs, trying to make it from submerged sheds to dry ground.

Bill Halemeyer and Lee Gettings, two of the neighboring farmers who helped Klunk save his grain, could legitimately say they had nothing better to do. Their own cash crop lay off in the distance toward Nutwood —under 20 feet of water.

Klunk fired up the truck and headed for the grain elevator.

Like the drought that struck the region in 1990, the floods caused widespread agricultural damage. This corn crop was a total loss.

The corn he was carrying would be just about the last grain carried to the elevator this year from the fertile Nutwood Drainage and Levee District.

And it was 1992's crop.

Not far away, Carney, who had been fighting the flood up in Quincy, Ill., before picking up the battle in Nutwood, awaited his next assignment.

"We're going to fight it all the way to Cairo (Ill.)," he gamely said, "but we're getting tired of losing."

Dusk was only minutes away when he gazed off to the west across the flood water toward the bluffs separating Illinois and Missouri.

"It looks like a beautiful sunset on a lake," he mused to no one in particular, "but there are people's homes under there."

8. As noted in the weather disaster map on page 36, the 1993 flood of the Mississippi River resulted in $26.7 billion in damage. What type of damage do you think contributed the most to these financial losses?

9. How might a drought or flood that occurs outside your area affect your daily life?

Global precipitation patterns

Investigation 2.2

Precipitation — any form of water that condenses or freezes from the atmosphere and that forms on or falls to Earth's surface. Forms of precipitation include rain, hail, sleet, snow, dew, and frost.

Aquifer — water-bearing rock formation.

Transpiration — the process of giving off water vapor through the skin (animals) or leaves (plants).

The atmosphere is one of the smallest global water reservoirs, yet it is one of the most important sources of water. Precipitation provides the water that nourishes plants and animals, feeds our rivers and lakes, and replenishes the groundwater stored in *aquifers*. Atmospheric water is the most easily renewed reservoir. A typical water molecule spends an average of nine days in the atmosphere before it returns to Earth's surface. In this unit, you will investigate global and regional precipitation patterns, runoff, evaporation, and *transpiration* to understand where precipitation falls and how it flows.

Precipitation and latitude

Aspects of weather, such as the amount and distribution of precipitation, vary considerably over short periods. Climate patterns appear at global, regional, and local scales when precipitation data are averaged over many decades. In this investigation, you will explore these patterns.

🖥 Launch ArcMap, then locate and open the **ewr_unit_2.mxd** file.

Refer to the tear-out Quick Reference Sheet located in the Introduction to this module for GIS definitions and instructions on how to perform tasks.

🖥 In the Table of Contents, right-click the **Global Precipitation Patterns** data frame and choose Activate.

🖥 Expand the **Global Precipitation Patterns** data frame.

The **Annual Precipitation** layer shows the annual global precipitation in centimeters per year.

1. Using data in the global precipitation layer legend, determine the range of mean annual precipitation on Earth. Be sure to include the units of measure, and round to the nearest whole number.

   a. Highest.

   b. Lowest.

2. Near what latitude is the mean precipitation the greatest?

3. In which latitude band in both the Northern and Southern Hemisphere (0° – 30°, 30° – 60°, 60° – 90°) is the mean precipitation the lowest?

🖥 Turn on the **Prime Meridian** layer.

Latitude and longitude

Latitude lines run east and west. Latitude is the angular distance on Earth's surface measured north or south from the equator, ranging from 0° at the equator to 90° at the poles.

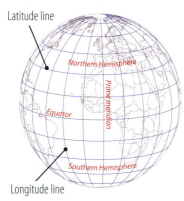

Longitude lines run north and south. Longitude is the angular distance measured east or west from the prime meridian, ranging from 0° at the prime meridian to 180° at the International Date Line.

The prime meridian is another name for the 0° longitude line. Examine the precipitation patterns along the prime meridian, both north and south of the equator.

4. How does precipitation in the Northern Hemisphere vary along the prime meridian between the following latitudes:

   a. Between the equator and 30° N?

   b. Between 30° N and 60° N?

   c. Between 60° N and 90° N?

5. How does precipitation in the Southern Hemisphere vary along the prime meridian between the following latitudes:

   a. Between the equator and 30° S?

   b. Between 30° S and 60° S?

   c. Between 60° S and 90° S?

Distribution of sunlight

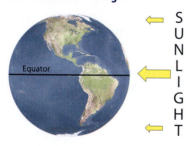

6. Are precipitation patterns the same in both hemispheres moving northward or southward away from the equator? If not, describe any differences you see.

Precipitation requires two things: water vapor and cooling. Most water vapor enters the atmosphere through the process of *evaporation*, and most cooling occurs as warm air expands and rises in a process called *convection*.

Global precipitation patterns are largely due to uneven solar heating of Earth's surface. In theory, the heaviest precipitation should be near the equator, where the Earth receives the most sunlight. There, the surface heating produces intense evaporation and convection.

Distribution of land

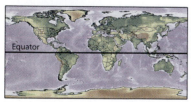

Northern Hemisphere
- 39% Land
- 61% Ocean

Southern Hemisphere
- 18% Land
- 82% Ocean

The distribution of land also affects precipitation patterns, because the sun heats land more easily than it heats the ocean. The equator and the Southern Hemisphere are mostly ocean, whereas the Northern Hemisphere contains large land masses. These land masses increase atmospheric heating north of the equator. As a result, the zone of highest precipitation is also shifted north of the equator.

Summarizing precipitation by latitude

Now you will quantify these global precipitation patterns. By creating a summary table, you will determine the mean precipitation for each 10-degree interval of latitude and graph the results. The summary table will combine data from the same latitudes in both hemispheres, helping to simplify your analysis of the precipitation patterns.

  💻 Click the Summarize button ⌊Σ⌋.

  💻 In the Summarize window, select **Annual Precipitation** as the feature layer.

  💻 Select **Latitude Range** as the field to summarize in the drop-down menu.

  💻 Double-click **Precipitation (cm/yr)** to display the statistics options and check **Average**.

  💻 Click **OK.**

A new window called a *summary table* will open. The resulting summary table should contain nine rows, one for each 10-degree latitude band, and four columns: OID, LATRANGE1, Count_ LATRANGE1, and Average_PRECIP_PER.

7. Complete Graph 1 using the mean annual precipitation (the column labeled "Average PRECIP PER") values for each 10-degree latitude band . Draw a bar graph by plotting the point halfway between the two latitudes, then filling in the area between the two latitudes. The value for the 10-degree latitude band has been plotted for you.

Graph 1 — Precipitation (cm/yr) versus latitude

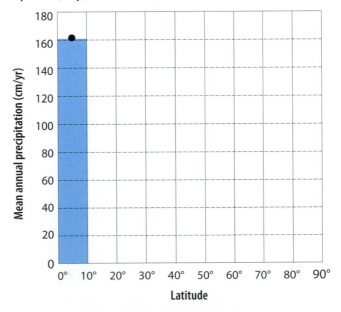

8. Re-read your response to questions 4 and 5, and study your completed graph. Do your answers to questions 4 and 5 agree with the precipitation data you graphed? Explain.

🖥 Close the summary table.

Precipitation in deserts and rain forests

Desert — a region that typically receives less than 25 cm (10 in) of precipitation per year. It also has evaporation rates that exceed the amount of precipitation it receives.

Next, you will look at the relationship between precipitation, latitude, and the global distribution of two of Earth's important ecosystems, deserts and tropical rain forests.

🖥 Turn on the **Deserts** and **Tropical Rain Forests** layers.

These layers show the locations of major world deserts and tropical rain forests.

9. On the globe diagram provided below

    • Mark the latitudes where deserts and tropical rain forests are found, using a different color or pattern for each ecosystem.

    • Complete the legend by filling in the color or pattern for each type of ecosystem.

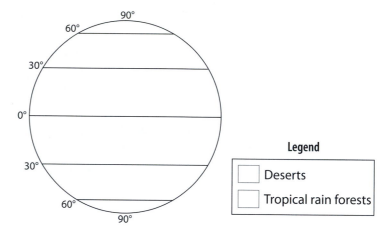

10. At what latitudes are most of the following features located:

    a. Deserts?

    b. Tropical rain forests?

11. How do the locations of deserts and tropical rain forests compare to the graph of global precipitation and latitude (Graph 1) on page 51?

12. What other latitudes receive very little precipitation, but are not identified on the map as desert regions?

13. Are there any major landmasses near the latitudes you identified in question 12? If so, what are they?

Global air circulation

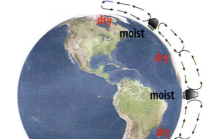

💻 Turn on the **Polar Arid Regions** layer.

14. Are these polar arid regions the same areas you indicated in question 13? If so, why aren't they considered deserts?

Rain forests are found near the equator where warm, moist air rises, cools, and condenses, producing precipitation. In the upper atmosphere, this air—now very dry—spreads out toward the poles. At around 30° north and south latitude, the cool, dry air sinks back to the surface, forming deserts on land. As you can see in the diagram at the left, this pattern repeats itself. Further north and south, relatively moist air is also found at around 60° north and south latitude, and dry air at the poles.

Precipitation in the deserts

You may be wondering, "How much more precipitation falls in the tropical rain forests than in the deserts?" Next, you will determine the mean precipitation received by these ecosystems each year.

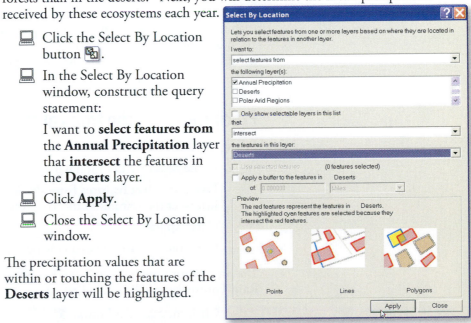

💻 Click the Select By Location button 🖳.

💻 In the Select By Location window, construct the query statement:

I want to **select features from** the **Annual Precipitation** layer that **intersect** the features in the **Deserts** layer.

💻 Click **Apply**.

💻 Close the Select By Location window.

The precipitation values that are within or touching the features of the **Deserts** layer will be highlighted.

💻 Click the Statistics button ⊠.

💻 In the Statistics window, calculate statistics for **only selected features** of the **Annual Precipitation** layer, using the **Precipitation (cm/yr)** field.

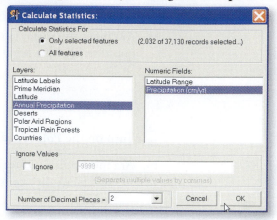

💻 Click **OK**. Be patient while the statistics are calculated.

15. Record the mean, minimum, and maximum precipitation values for the deserts in Table 1.

Table 1 — Precipitation in desert and rain-forest ecosystems

Region	Mean precipitation cm/yr	Minimum precipitation cm/yr	Maximum precipitation cm/yr
Deserts			
Tropical rain forests			
Global			

💻 Close the Statistics window.

According to Graph 1 on page 51, the precipitation at 30° latitude is much higher than the maximum value for the deserts at these latitudes in Table 1. In fact, not all land at these latitudes is desert. For example, the southeastern U.S. is at this latitude (30° N), yet receives considerable precipitation due to its proximity to warm ocean currents.

Precipitation in the tropical rain forest

💻 Click the Select By Location button 🖳.

💻 In the Select By Location window, construct the query statement:

I want to **select features from** the **Annual Precipitation** layer that **intersect** the features in the **Tropical Rain Forests** layer.

💻 Click **Apply**.

💻 Close the Select By Location window.

The precipitation values that are within or touching the features of the **Tropical Rain Forests** layer will be highlighted.

💻 Click the Statistics button ⊠.

- In the Statistics window, calculate statistics for **only selected features** of the **Annual Precipitation** layer, using the **Precipitation (cm/yr)** field.
- Click **OK**.

16. Record the mean, minimum, and maximum precipitation values for the tropical rain forests in Table 1.

- Close the Statistics window.

Mean global precipitation

For comparison, determine the global mean annual precipitation.

- Click the Clear Selected Features button.
- Click the Statistics button.
- In the Statistics window, calculate statistics for **all features** of the **Annual Precipitation** layer, using the **Precipitation (cm/yr)** field.
- Click **OK**.

17. Record the mean global precipitation in Table 1.

At left is a list of mean annual precipitation figures for several U.S. cities.

18. In terms of mean annual precipitation, which two cities in the U.S. are most similar to deserts?

a.

b.

19. Which two cities are most similar to rain forests?

a.

b.

20. Why do you think these cities receive so much precipitation?

Mean annual precipitation for selected U.S. cities (cm/yr)

Baton Rouge, Louisiana	165.4
Boston, Massachusetts	111.3
Charlotte, North Carolina	109.7
Denver, Colorado	38.9
Hilo, Hawaii	327.4
Los Angeles, California	30.7
Nashville, Tennessee	123.2
New York City, New York	108.8
Salt Lake City, Utah	38.9
St. Louis, Missouri	86.1
Tucson, Arizona	28.2
Seattle, Washington	98.0

- Close the Statistics window.
- Quit ArcMap and do not save changes.

Reading 2.3

Moving air and water

Where the wind blows

Winds are caused by differences in temperature, which result in differences in air pressure. Where air is warm, it expands and rises, creating low pressure. Where air is cool, it contracts and sinks, creating high pressure. Generally, winds blow from areas of high pressure to areas of low pressure as the atmosphere acts to equalize the pressure differences.

Global surface winds

Uneven heating of Earth's surface by the sun and the force generated by Earth's rotation combine to create wind belts that encircle the planet. Near the equator, where the sun's rays strike the surface most directly, the air is heated, causing it to expand and rise. This forms a low-pressure belt around the Earth at the equator. At the Earth's surface, cooler air north and south of the equator flows toward the low-pressure belt to replace the rising air. The **Coriolis effect**, caused by Earth's rotation, deflects these winds to the right in the Northern Hemisphere and to the left in the Southern Hemisphere. Thus, rather than blowing directly toward the equator, the winds blow from east to west in both hemispheres. These easterly winds are called the **trade winds**.

The rising air at the equator condenses and precipitates much of its moisture, creating the tropical rain forests. As the air rises, it travels toward the poles, eventually cooling, drying out, and becoming more dense. At about 30° north and south latitude, this cool, dense air sinks, forming a band of high pressure that is responsible for the deserts found at these latitudes. When this sinking air reaches the surface, some of it spreads toward the poles. This air is again deflected by the Coriolis effect, but this time the winds blow from west to east. These winds are known as the **prevailing westerlies** (Figure 1).

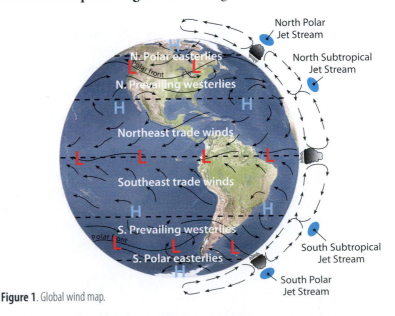

Figure 1. Global wind map.

At the poles, cold, sinking air creates a high-pressure area. As the sinking air moves toward the equator, it is deflected toward the west, forming the wind belt called the **polar easterlies**. At around 60° latitude, this air warms and rises, forming a weak low-pressure zone. This is where most winter storms begin.

Air masses and fronts

Air masses are large volumes of air that remain in one place long enough to acquire the humidity and temperature characteristics of the surface beneath them. Most weather is caused by interactions between air masses in the lower atmosphere, which extends from Earth's surface to an altitude of about 8 km at the Earth's poles and 16 km above the equator.

Continental air masses, abbreviated *c*, form over land and are generally dry. Maritime air masses, abbreviated *m*, form over oceans and are generally moist. Tropical air masses, abbreviated *T*, form near the equator and are typically warm. Polar air masses, abbreviated *P*, form at high latitudes and are cold. The coldest air masses form near the poles, and are called arctic air masses, abbreviated *A*.

Combining these characteristics defines six main types of global air masses, as shown in Figure 2.

- cT — continental tropical
- cP — continental polar
- cA — continental arctic
- mT — maritime tropical
- mP — maritime polar
- mA — maritime arctic

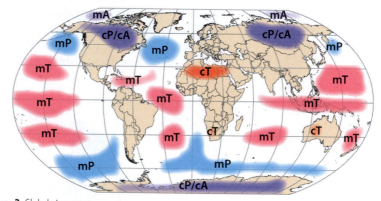

Figure 2. Global air masses.

The boundaries between air masses are called **fronts**. The temperature and humidity differences along these frontal boundaries produce clouds, precipitation, wind, and other weather phenomena. The three basic types of fronts — cold fronts, warm fronts, and stationary fronts — are illustrated in Figure 3 on the following page.

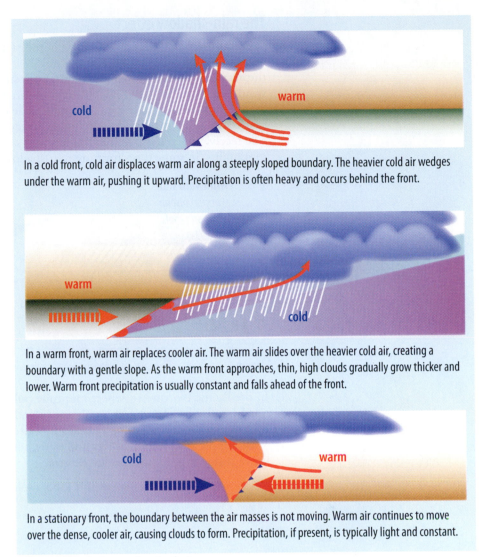

In a cold front, cold air displaces warm air along a steeply sloped boundary. The heavier cold air wedges under the warm air, pushing it upward. Precipitation is often heavy and occurs behind the front.

In a warm front, warm air replaces cooler air. The warm air slides over the heavier cold air, creating a boundary with a gentle slope. As the warm front approaches, thin, high clouds gradually grow thicker and lower. Warm front precipitation is usually constant and falls ahead of the front.

In a stationary front, the boundary between the air masses is not moving. Warm air continues to move over the dense, cooler air, causing clouds to form. Precipitation, if present, is typically light and constant.

Figure 3. Three basic types of weather fronts.

Precipitation, air masses, and topography

For rain or snow to occur, two conditions must be met. First, the lower atmosphere must have large quantities of water vapor, or precipitable water. Second, the air mass must cool sufficiently. Cooling occurs when an air mass is forced upward. There are four types of uplift:

- convective uplift—an air mass moves over a warmer surface, expands, and rises.
- frontal uplift—a denser, cooler air mass "wedges" underneath a warmer air mass and forces it upward (Figure 3).
- orographic uplift—an air mass is forced up and over a landform such as a mountain or high plateau (Figure 4 on the following page).
- disturbance uplift—conditions in the upper atmosphere, often associated with the jet stream, produce uplift in the lower atmosphere.

The rain-shadow effect

Orographic uplift occurs when warm, moist air is pushed up and over a mountain range (Figure 4). The air cools as it rises, causing the water vapor in the air to condense and fall as either rain or snow on the **windward** (facing the wind) side of the mountains. Having lost much of its moisture, the air descends the **leeward** (opposite of wind direction) side of the mountain range, compresses, and warms, creating a warm, dry wind. Valleys and lowlands on the leeward side receive much less precipitation and are said to be in the rain shadow of the mountains, in some cases forming rain-shadow deserts.

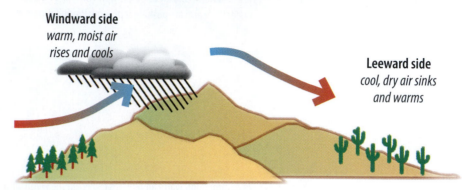

Windward side
warm, moist air rises and cools

Leeward side
cool, dry air sinks and warms

Figure 4. Orographic uplift creates rain-shadow deserts on the leeward slopes of mountain ranges.

In the Pacific Northwest, where maritime polar (mP) air masses from the northern Pacific Ocean carry moisture inland, the north-south running Cascade Mountains produce a strong rain-shadow effect. Although most people think of Washington and Oregon as wet and rainy, the eastern halves of those states on the leeward side of the Cascades are deserts.

1. Explain how air masses and topography interact to produce rain shadows.

Upper level winds — the jet streams

During World War II, bomber pilots flying at altitudes above six kilometers confirmed the existence of upper-level air currents predicted by atmospheric scientists to blow at speeds of nearly 300 km/hr (approximately 190 mi/hr). These high-altitude "rivers of air," called jet streams, play an important role in the formation and movement of storms.

Although television meteorologists usually refer to a single "jet stream," there are actually two main jet streams in each hemisphere. These are the **polar jet stream** and the **subtropical jet stream**, and they encircle the globe along winding paths thousands of kilometers long and hundreds of kilometers wide.

The polar jet stream flows at an altitude of 5–12 km along the polar front—the boundary between cold, dry high-latitude air and warm, moist middle-latitude air. The polar jet stream influences precipitation in the U.S. by contributing to the formation and movement of storms along the polar front.

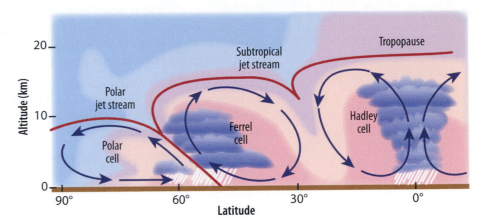

Figure 5. Cross section of Earth's atmosphere, showing locations of the jet streams.

The subtropical jet stream flows at an altitude of 8–15 km above the boundary between cool mid-latitude air and warm tropical air. The subtropical jet stream contributes moisture to summer storms in the southern U.S., but is much weaker and has less influence on weather patterns than the polar jet stream.

2. Which jet stream has the most influence on precipitation in the U.S.? Explain.

Where the water flows

The interaction between air masses, global winds, and topography determine where precipitation occurs. When precipitation reaches the ground, gravity continues to pull the water downward. The characteristics of Earth's surface influence the water's path. If the surface is **permeable** (permits water to pass through) and **porous** (contains empty spaces) water can move through or infiltrate it.

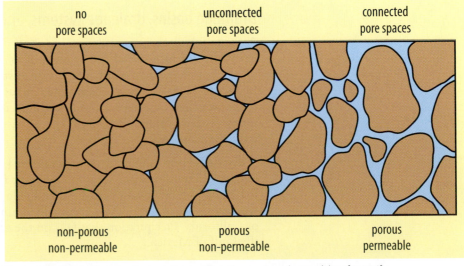

Figure 6. Illustration of how the alignment of grains affects the porosity and permeability of an aquifer.

If Earth were flat and featureless, permeability and porosity would completely dictate water flow. However, Earth has mountains, hills, and valleys, so water flow is determined by the interaction between the surface properties (porosity and permeability, for example) and surface features (such as topography).

Water in the soil can return directly back to the atmosphere through evaporation; it can be taken up by plant roots and returned to the atmosphere by transpiration; or it can continue infiltrating downward to join groundwater in a water-bearing rock formation, or **aquifer**.

3. How do permeability and porosity affect the amount of precipitation that infiltrates the ground?

Topography: slope and aspect

Topography is the shape or *configuration* of a land surface that shows relative elevations and positions of natural and man-made features. Two facets of topography, **slope** and **aspect**, affect water flow. Slope is the steepness of the surface, expressed as an angle. Slope influences the *speed* that water flows downhill. For example, water moving down a mountain with a 40° slope flows faster than water flowing down a 10° slope. Aspect describes the *direction* that the slope faces — north, south, east, or west — and determines the direction that water flows on the surface.

4. How do slope and aspect affect water flow?

Drainage basins, drainage systems, and divides

Aspect controls the accumulation of runoff from precipitation. A **drainage basin** is an area in which runoff flows downhill to a common point, such as a lake, or to a common channel, such as a river or stream, as shown in Figure 7 on the following page. Within the basin, the land aspect faces inward. A drainage basin is characterized by a network of interconnected creeks, streams, and rivers that merge with a common **drainage**. The Mississippi River Basin is an example of a large drainage basin.

A **divide** is the boundary between drainage basins. At a divide, the opposing aspects of the land cause water to flow away from the boundary (Figure 7). The Rocky Mountains are an example of a divide. Precipitation that falls on the western slopes of the Rockies eventually drains into the Pacific Ocean or the Gulf of California. Precipitation that falls on the eastern slopes eventually drains into the Gulf of Mexico.

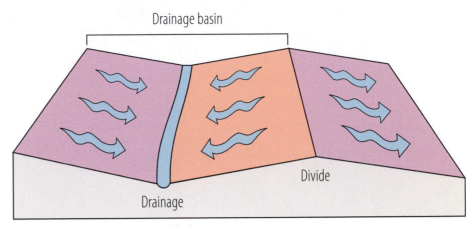

Figure 7. A drainage basin, drainage, and divide.

Watersheds

Drainage basins and divides define larger regions known as **watersheds**. A watershed is the total area drained by a river and its tributaries. Figure 8 shows the Colorado River watershed, a large, multi-state region that includes many drainage basins that eventually drain into the Colorado River. The term watershed is often used interchangeably with the term drainage basin, in that it can also be applied to relatively small areas. For example, the land that drains into a small stream may also be called a watershed.

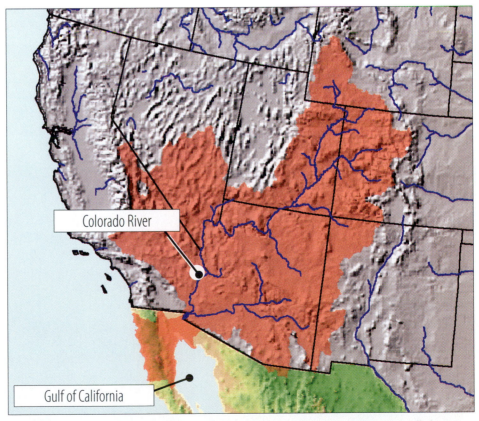

Figure 8. The shaded area represents the Colorado River watershed (or basin). Water from this region eventually drains to the Colorado River and the Gulf of California.

5. What is the difference between a drainage basin and a divide?

Investigation 2.4A

U.S. precipitation patterns

In this investigation, you will examine the precipitation patterns that characterize the climate of the United States. You will find that latitude is just one factor among many in the distribution, quantity, and timing of precipitation.

East versus West—regional precipitation

First, you will explore precipitation patterns in the United States, paying particular attention to the relationship between geography and precipitation.

💻 Launch ArcMap, then locate and open the **ewr_unit_2.mxd** file.

Refer to the tear-out Quick Reference Sheet located in the Introduction to this module for GIS definitions and instructions on how to perform tasks.

💻 In the Table of Contents, right-click the **U.S. Precipitation Patterns** data frame and choose Activate.

💻 Expand the **U.S. Precipitation Patterns** data frame.

The **U.S. Precipitation** layer shows the mean annual precipitation for the U.S. in centimeters per year. Blues and purples represent the highest precipitation amounts, and reds and browns represent the lowest. Notice how dramatically the precipitation varies across the country, particularly in the Western states.

1. Which regions of the U.S. receive the most precipitation?

Regions of the U.S.

2. In areas with high precipitation, what generally happens to the amount of precipitation when moving farther inland from the ocean? (Does it increase or does it decrease?)

Clearly, the distance from a large water source such as an ocean is a major factor in the amount of precipitation in a region.

The 100th meridian

In 1879, the famous explorer and scientist John Wesley Powell noted that a natural dividing line based on precipitation seemed to separate the eastern and western U.S. This line is near the 100th meridian (100° West longitude). Next, you will quantify the differences between these regions.

💻 Turn on the **100th Meridian** layer. It appears as a heavy purple line that cuts the U.S. roughly in half.

West of the 100th meridian

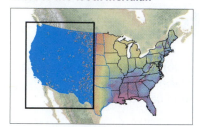

When you select the area west of the 100th meridian (the purple line), try to select as close to the line as possible without going past it.

🖥 Select the **U.S. Precipitation** layer.

🖥 Use the Select Features tool 🔲 to drag a box that encloses all of the area west of the 100th meridian (see sidebar). The selected region will be highlighted on your map.

🖥 Click the Statistics button 🗵.

🖥 In the Statistics window, calculate statistics for **only selected features** of the **U.S. Precipitation** layer, using the **Precipitation (cm/yr)** field.

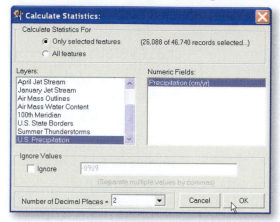

🖥 Click **OK**. Be patient while the statistics are calculated.

3. Record the mean, minimum, and maximum precipitation values for the western U.S. (**West of 100° W**) in Table 1.

Table 1 — Comparing precipitation east and west of the 100th meridian

	West of 100° W	East of 100° W
Mean precipitation (cm/yr)		
Minimum precipitation (cm/yr)		
Maximum precipitation (cm/yr)		

🖥 Close the Statistics window.

🖥 Click the Switch Selected Features button 🗺 to select the region east of the 100th meridian.

🖥 Repeat the statistics procedure above to calculate precipitation statistics for the eastern U.S. and record them in Table 1.

4. Which side of the U.S. (east or west) has

  a. the higher mean annual precipitation?

  b. the larger range of precipitation values?

5. What do you think accounts for the difference in precipitation patterns between the eastern and western U.S.? Explain.

🖥 Close the Statistics window.

🖥 Turn off the **100th Meridian** layer.

🖥 Click the Clear Selected Features button ▣ to deselect the eastern states.

Precipitation and air masses

After exploring some of the general patterns of precipitation across the U.S., it is time to consider some explanations for the patterns you have observed. For precipitation to occur, the lower atmosphere must have large quantities of water vapor, or precipitable water. The regions where air masses originate affect the amount of precipitable water they are likely to contain.

Precipitable water — the amount of water that would be produced on the surface if all of the water vapor over an area were condensed to a liquid.

🖥 Turn on the **Air Mass Sources** layer group.

🖥 Select the **Air Mass Sources** layer group.

🖥 Click the Full Extent button 🌐 to show the entire air mass layer group.

Air mass — a large volume of air with a fairly uniform temperature, pressure, and moisture content.

The colors of the **Air Mass Sources** layer group represent the amount of precipitable water in each air mass (in kilograms of water per square meter). Examine the source regions (maritime or continental; and arctic, polar, or tropical) and relate the amount of precipitable water to the source region's climate.

6. Use the color spectrum below to rank these five air mass types from driest to wettest: cA, mP, cP, mT, and cT. The first one (cA) has been done for you. The maritime arctic (mA) air mass is not shown on the map.

Driest   **Wettest**

cA

7. Which air masses hold more precipitable water — those that form over continents, or those forming over oceans? Why?

8. Which air masses hold more precipitable water — those that form in tropical regions, in polar regions, or in arctic regions? Why?

🖥 Turn off the **Air Mass Sources** layer group.

🖥 Use the Zoom In tool 🔍 to zoom in on the U.S. again.

Precipitation and topography—orographic uplift

When an air mass encounters mountains, it is pushed upward. If the air mass contains enough moisture and is cooled sufficiently, precipitation will occur. As you will recall from the reading, this phenomenon is called orographic uplift. Now you will investigate how topography affects precipitation, both regionally and locally.

🖥 Turn on the **U.S. Topography** layer.

The **U.S. Topography** layer is a shaded relief map of the United States, with major mountain ranges and basins visible as wrinkles in the map. Areas with little topographic variation appear smooth, almost featureless. Look at the patterns across the U.S. and answer the following question.

9. Compare the topography in the western U.S. to the topography of the eastern U.S. (For example, where is the land mountainous? Where is it flat?)

The **U.S. Precipitation** layer is semi-transparent, so you should be able to see through the layer to the underlying shaded relief image. This allows you to examine topography and precipitation together.

10. Describe the relationship between precipitation and topography. Compare the precipitation and topography of the eastern U.S. to the western U.S.

Precipitation in the Pacific Northwest

The Pacific Northwest includes the states of Washington and Oregon. This region is famous for its moist climate and lush rain forests.

11. Considering the location of the Pacific Northwest (latitude, proximity to the ocean), which type of air masses typically bring moisture to this region (mT, mP, mA, cT, cP, or cA)?

12. Locate and circle the Pacific Northwest region on the global wind map (Figure 1) on the following page.

Temperature and elevation

As air rises, atmospheric pressure decreases, causing the air to expand and cool. As air sinks, the pressure increases, compressing the air and causing its temperature to increase. The mean rate of temperature change with increasing altitude, called the *lapse rate*, is about −6.5 °C/km up to 11 km, above which the atmosphere remains at a constant −56.5 °C.

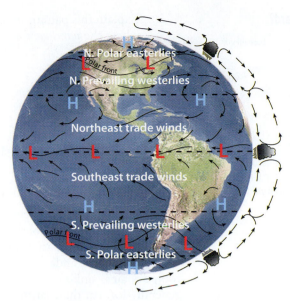

Figure 1. Global wind map.

13. According to Figure 1, where do the air masses that bring precipitation to the Pacific Northwest originate?

To explore the relationship between precipitation and elevation, you will examine two kinds of data profiles. A *topographic profile* shows local relief — mountain ranges and basins — as well as the overall elevation. A *precipitation profile* along the same line, or *transect,* will help you understand how precipitation relates to topography.

💻 Turn on and select the **Profiles** layer.

This layer shows two transects, one that extends from the state of Washington eastward into Idaho, the other from Louisiana northward into Illinois.

💻 Click on the Washington–Idaho transect using the Hyperlink tool ⚡.

In the window that opens, the upper profile shows annual precipitation along the transect line and the lower profile shows the elevation along the same transect. You may want to arrange the windows on your screen so you can see both the transect line in the data frame and the window containing the profiles.

14. According to the profiles, which mountains receive more precipitation — those near the coast, or those farther inland?

15. Within a mountain range, how does the precipitation on the western slope compare to the precipitation on the eastern slope?

Understanding profiles

A *profile* is a side view of a surface produced along a line called a *transect*.

- A topographic profile shows the elevation of the land along a transect.
- A precipitation profile shows the accumulated precipitation along a transect for a year. The peaks are the places that received the most precipitation, and the valleys are the places that received the least precipitation.

Windward or leeward?

Windward side   Leeward side

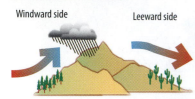

The side of a mountain that is facing the direction of the wind is called the *windward* side, where air rises, cools, and condenses. The air sinks, warms, and expands on the other side of the mountain, called the *leeward* side.

16. Explain this pattern in terms of orographic uplift. (Hint: Think about where the *windward* and *leeward* sides of each mountain range are located.)

🖥 Close the Washington–Idaho profile window.

Precipitation in the Southeast

Now, you will examine precipitation and topography patterns in the southeastern U.S. Look again at the global wind map on page 69. Notice that the southeastern U.S. is near the boundary between the easterly trade winds and the prevailing westerlies. This boundary is identified by a global band of high pressure, where cool, dry air is sinking toward the surface. As you saw earlier, the world's deserts generally form at these latitudes, including the deserts of the southwestern U.S. The southeastern U.S., although at the same latitude, is certainly not a desert.

17. Speculate on the direction or geographic location from which the moisture that produces precipitation in the southeastern U.S. originates.

18. Which type of air mass typically brings moisture to this region (mT, mP, mA, cT, cP, or cA)?

🖥 Click on the Louisiana–Illinois transect using the Hyperlink tool ⚡.

These profiles show the precipitation and topography along a transect from Louisiana to southern Illinois. For easy comparison, the vertical scales of the profiles are the same as those of the Washington–Idaho profiles.

19. Use the profiles to describe the topography of the states crossed by the transect. (Are they mountainous, flat, hilly?)

20. Is the amount of precipitation along this transect influenced more by topography or by distance from the coastline? Explain.

🖵 Close the Louisiana–Illinois profile window.

🖵 Turn off the **U.S. Precipitation** and **Profiles** layers.

21. On the basis of your observations, describe how the movement of air masses, distance from the ocean, and topography affect the amount of precipitation a location receives.

🖵 Quit ArcMap and do not save changes.

Investigation 2.4B

U.S. precipitation patterns

In this investigation, you will examine how the jet stream and convection affect the precipitation patterns of the United States. You will discover that there are many factors, aside from latitude and air masses, that affect these patterns.

Precipitation and the polar jet stream— frontal and disturbance uplift

You may have seen or heard of something called the jet stream during TV weather reports. The jet stream is a high speed current of air embedded in the upper troposphere, at an altitude of about 9 kilometers (30,000 ft). The jet stream marks the boundary between the warm tropical air and cold polar air.

Troposphere— the layer of the atmosphere that is closest to earth, and extends to an altitude of 11 - 16 kilometers. Convection is active and most clouds form in the troposphere.

Tropopause— the upper boundary of the troposphere.

Figure 1. Location of the polar jet stream.

The position of the polar jet stream is strongly related to precipitation patterns in the U.S. The polar jet stream flows above the polar front—the boundary between cold, dry air to the north and warm, moist air to the south—and contributes to the formation of frontal storms that bring substantial amounts of precipitation to the northern and central U.S., particularly during the winter.

🖥 Launch ArcMap, then locate and open the **ewr_unit_2.mxd** file.

Refer to the tear-out Quick Reference Sheet located in the Introduction to this module for GIS definitions and instructions on how to perform tasks.

🖥 In the Table of Contents, right-click the **U.S. Precipitation Patterns** data frame and choose Activate.

🖥 Expand the **U.S. Precipitation Patterns** data frame.

🖥 Click the Media Viewer button 🔣 and open the **Jet Stream Movie**. The position of the jet stream is represented by the fastest windspeeds (reds, oranges, and yellows).

Jet Stream Movie

If you have difficulty interpreting the animation, this data frame also includes layers showing the average position and wind speed of the jet stream for the months of January, April, July, and October. Keep in mind that this is an average position — the actual position of the jet stream is constantly changing.

This animation shows the changing position of the jet stream over the course of a year, averaged from 30 years of data. The month is shown in the bottom right of the Movie window. Watch the movie several times, noting the position and speed of the jet stream at different times of the year. The jet stream always flows west to east, but its latitude varies with the seasons. The colors represent the wind speed, with reds and oranges representing the fastest winds (the jet stream). On the basis of the movie, answer the following questions.

1. During which season is the jet stream farthest north?

2. During which season is the jet stream farthest south?

3. During which season is the speed of the jet stream the fastest? And in which season is it slowest?

   a. Fastest.

   b. Slowest.

🖥 Close the Media Viewer window.

Precipitation and thermal convection—convective uplift

The jet stream does not explain all precipitation patterns in the U.S. The final piece of the precipitation puzzle are the thunderstorms that rumble and roll throughout much of the country each summer. Thunderstorms develop when intense heating at the surface causes *thermal convection*. The heated air rapidly moves upward and cools, and the water vapor precipitates out. Thunderstorms provide a substantial amount of the annual precipitation for some regions of the country.

🖥 Turn on the **Summer Thunderstorms** layer.

This layer shows the mean number of days in which thunderstorms occurred during July, August, and September.

4. In which two parts of the country are summer thunderstorms most common?

   a.

   b.

🖥 Turn on the **Air Mass Sources** layer group.

Watch a summer thunderstorm!

To view time-lapse movies of summer thunderstorms in Arizona, click the Media Viewer button 🎞 and choose **Thunderstorm Movie 1**, **2**, or **3**. Do not spend too much time viewing the movies — you still have data to analyze.

5. Which type(s) of air masses provide the moisture that feeds the thunderstorms in each of these regions?

Now you will take a closer look at one of these areas.

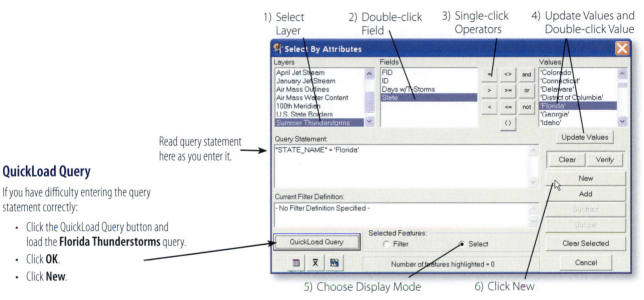

- Click the Select By Attributes button.
- To more closely examine the summer thunderstorms in Florida, query the **Summer Thunderstorms** layer for **("State" = 'Florida')** as shown in steps 1–6. Your query will actually read:

"STATE_NAME" = 'Florida'

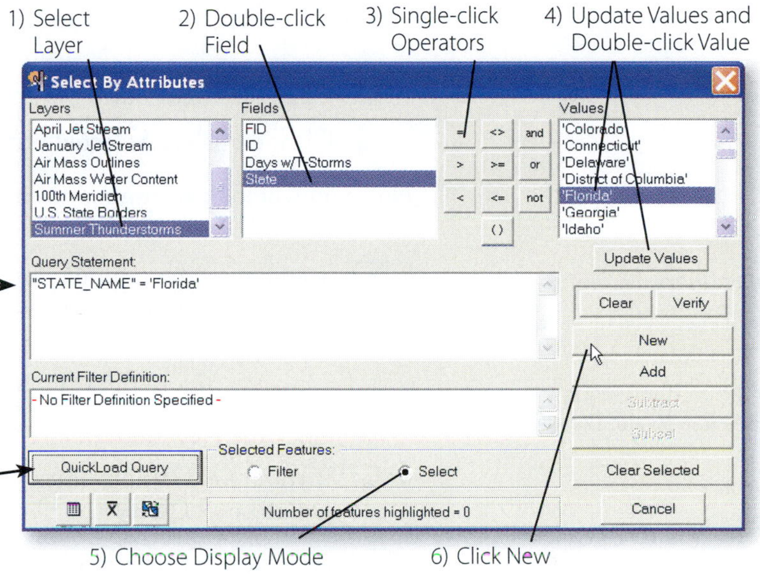
1) Select Layer
2) Double-click Field
3) Single-click Operators
4) Update Values and Double-click Value
Select By Attributes
Layers
April Jet Stream
January Jet Stream
Air Mass Outlines
Air Mass Water Content
100th Meridien
U.S. State Borders
Summer Thunderstorms
Fields
FID
ID
Days w/T-Storms
State
Values
'Colorado'
'Connecticut'
'Delaware'
'District of Columbia'
'Florida'
'Georgia'
'Idaho'
Update Values
Query Statement:
"STATE_NAME" = 'Florida'
Clear
Verify
New
Add
Current Filter Definition:
- No Filter Definition Specified -
QuickLoad Query
Selected Features:
Filter
Select
Clear Selected
Number of features highlighted = 0
Cancel
5) Choose Display Mode
6) Click New

Read query statement here as you enter it.

QuickLoad Query

If you have difficulty entering the query statement correctly:

- Click the QuickLoad Query button and load the **Florida Thunderstorms** query.
- Click **OK**.
- Click **New**.

- If you have difficulty entering the query statement correctly, refer to the **QuickLoad Query** described at left.

The state of Florida should now be highlighted on your map.

- Click the Statistics button in the Select By Attributes window.
- In the Statistics window, calculate statistics for **only selected features** of the **Summer Thunderstorms** layer, using the **Days w/T-Storms** field.

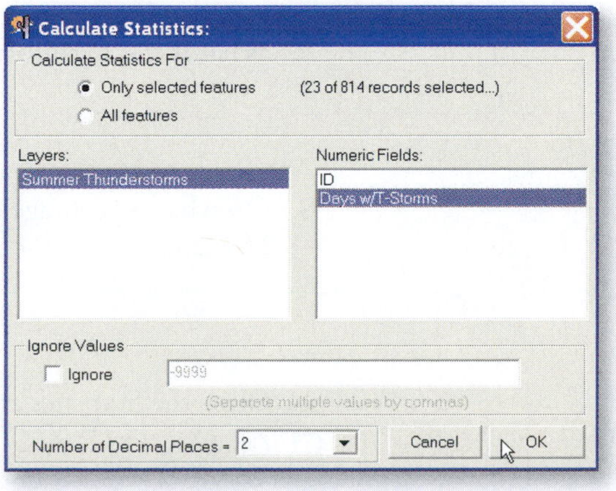
Calculate Statistics:
Calculate Statistics For
Only selected features
(23 of 814 records selected...)
All features
Layers:
Summer Thunderstorms
Numeric Fields:
ID
Days w/T-Storms
Ignore Values
Ignore
-9999
Number of Decimal Places = 2
Cancel
OK

- Click **OK**. Be patient while the statistics are being calculated.

The mean number of thunderstorms Florida has each year is given as the **Mean**.

6. What is the mean number of summer thunderstorm days in Florida? Round to the nearest whole number.

🖥 Close the Statistics window.

How many days?

"Thirty days hath September...", and July and August both have 31 days; so the summer data cover a total of 92 days.

7. On what *percentage* of summer days do thunderstorms occur in Florida? (Hint: Divide the mean number of thunderstorms in question 6 by the number of days in July through September, multiply the result by 100, and round to the nearest whole number.)

🖥 Click the **Clear** button in the Select By Attributes window.

🖥 Click the **Clear Selected** button in the Select By Attributes window.

Now you will look at summer thunderstorms in New Mexico, in the southwestern U.S.

QuickLoad Query

If you have difficulty entering the query statement correctly:

- Click the QuickLoad Query button and load the **New Mexico Thunderstorms** query.
- Click **OK**.
- Click **New**.

🖥 To examine the summer thunderstorms in New Mexico, query the **Summer Thunderstorms** layer for (**"State"** = **'New Mexico'**). The query will actually read:

"STATE_NAME" = 'New Mexico'

🖥 Click the **New** button.

🖥 If you have difficulty entering the query statement correctly, refer to the **QuickLoad Query** described at left.

The state of New Mexico should now be highlighted on your map.

🖥 Click the Statistics button ⊠ in the Select By Attributes window.

🖥 In the Statistics window, calculate statistics for **only selected features** of the **Summer Thunderstorms** layer, using the **Days w/T-Storms** field.

🖥 Click **OK**. Be patient while the statistics are being calculated.

The mean number of thunderstorms New Mexico has each year is given as the **Mean**.

8. What is the mean number of summer thunderstorm days in New Mexico?

9. On what percentage of summer days do thunderstorms occur in New Mexico?

🖥 Close the Statistics and Select By Attributes windows.

🖥 Click the Clear Selected Features button ⊠.

To determine how important thunderstorms are to the Southwest and Southeast, you will view a movie that shows the mean percentage of the annual precipitation that each state receives in each month.

🖵 Click the Media Viewer button 📖 and open the **U.S. Mean Precipitation Movie**. Play the movie several times and watch how precipitation varies across the country throughout the year. Use the movie's map legend to estimate the percentage of annual rainfall. You can use the Pause button to stop the movie and see specific months more clearly if necessary.

10. What percentage of its annual rainfall does New Mexico receive in each month from July through September? Use the highest percentage you see anywhere in New Mexico in the movie (in each of those three months) for your answer, then add them up for a total percentage of the state's annual rainfall.

    a. July

    b. August

    c. September

    d. Total percentage

11. What percentage of its annual rainfall does Florida receive in each month from July through September? Use the highest percentage you see anywhere in Florida in the movie (in each of those three months) for your answer, then add them up for a total percentage of the state's annual rainfall.

    a. July

    b. August

    c. September

    d. Total percentage

12. Where do more summer thunderstorms occur — in the southwestern or in the southeastern U.S.?

13. Where are summer thunderstorms more important to the annual precipitation — in the Southwest or in the Southeast? Explain your answer.

⌨ Close the Media Viewer window.

⌨ Quit ArcMap and do not save changes.

Investigation 2.5

Surface water flow

In this investigation, you will examine what happens to precipitation when it reaches Earth's surface.

Aspect, divides, and watersheds

To better understand surface water flow in the U.S., you will begin by looking at the *aspect*—the main direction in which the land slopes. Because water flows downhill along the steepest path, the aspect of the surface indicates the direction that runoff will flow.

🖥 Launch ArcMap, then locate and open the **ewr_unit_2.mxd** file.

Refer to the tear-out Quick Reference Sheet located in the Introduction to this module for GIS definitions and instructions on how to perform tasks.

🖥 In the Table of Contents, right-click the **Surface Water Runoff** data frame and choose Activate.

🖥 Expand the **Surface Water Runoff** data frame.

The **Land Aspect** layer uses color to show the general direction in which precipitation falling on a given land area will flow. The graphic at left shows the colors used to represent aspect, and the direction that each color represents. You will use this layer to determine the primary direction that water flows in each of four regions of the U.S.

🖥 Use the predominant color to determine the general direction that water flows in each region. (For example, if more of the region looks red than any other color, water flows toward the north.)

1. On Map 1 below, draw arrows showing the primary direction of water flow in each of the four major regions of the U.S.

Aspect directions

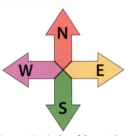

The colors used in the **Land Aspect** layer represent the general direction that runoff will flow in each region.

Hint for question 1

If you need help determining the flow direction, click the Media Viewer button 🖼 and open the **U.S. Land Aspect by Region** image. The graphs show the percentage of land in each region that slopes in each of the four directions (north, south, east, and west).

Map 1 — Direction of water flow in the U.S.

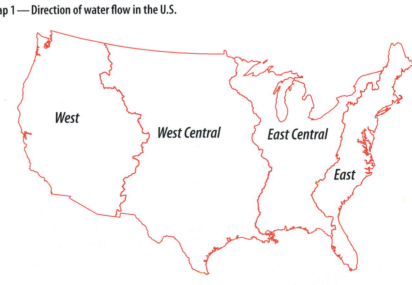

Drainages and divides

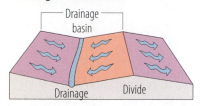

- Drainage basin
- Drainage
- Divide

Drainage basins are areas in which runoff flows downhill to a common point, such as a lake, or to a common channel. Divides form the boundaries between drainage basins.

2. On the basis of the arrows you drew, how many continental divides are shown on Map 1 on the previous page? Mark the divide(s) on the map using heavy dashed lines. (See **Drainages and divides**, at left.)

3. Shade in the largest drainage basin shown on Map 1. (Hint: Drainage basins are bounded by divides.)

4. On Map 1, use a heavy, solid line to represent the major river that drains this basin.

5. What is the name of this river system, the largest in the U.S.? (If necessary, consult a U.S. map or atlas.)

🖥 Turn off the **Land Aspect** and **Aspect Regions** layers.

Watersheds and river systems

🖥 Turn on the **Major U.S. Watersheds** layer.

The **Major U.S. Watersheds** layer shows the locations of the major river basins in the United States. This part of the investigation explores the six largest watersheds. To highlight these watersheds:

🖥 Select the **Major U.S. Watersheds** layer.

🖥 Click the Open Attribute Table button 🔲 to open the **Major U.S. Watersheds** attribute table.

🖥 Right-click on the **Area (km2)** field heading and choose Sort Descending to sort the features from largest to smallest area.

🖥 Holding the Ctrl key down, use the Pointer tool 🔺 to click and select the top six rows of the attribute table (on the left side of the table where the arrow is located). The rows should be highlighted, indicating that they have been selected.

🖥 Close the attribute table and examine the map. The six large watersheds you selected should be highlighted. If not, open the attribute table and repeat the selection process.

6. How many of the largest watersheds are completely within the United States?

Remember that a watershed is an area drained by a network of interconnected streams and rivers. Runoff and smaller streams flow into a common body of water —usually a larger stream or river. Therefore, everyone living downstream in a watershed is affected by what happens upstream. This includes both the quantity and quality of the water that they receive.

7. Choose one of the watersheds that is shared by the U.S. and a neighboring country (either Canada or Mexico). Explain how you think the job of managing this watershed might be complicated by the fact that it is shared by two countries.

Measuring stream discharge

Stream *discharge* is the volume of water moving past a point in a given length of time. To calculate discharge, you must know the depth and speed of the water. These are measured at a *gauging station* like the one shown below.

Gauging station at the outlet of the U.S. Department of Agriculture/Agricultural Research Service Walnut Gulch Experimental Watershed near Tombstone, Arizona. Shown here during a flash flood, this outlet drains an area covering 148 km² (57 mi²).

Is there a difference between runoff and discharge?

The terms *runoff* and *discharge* are often used interchangeably. Scientifically speaking, runoff from a watershed can be calculated by dividing the discharge at the outlet of a watershed (a volume of water per given length of time) by the area of the watershed. It is often expressed in units of meters per year or kilometers per year.

Kilometers to centimeters

1 km = 1,000 m

1 m = 100 cm

So, 1 km = 100,000 cm

River discharge and basin runoff

In this section, you will explore how runoff varies by region in each of the six largest U.S. watersheds. It is not possible to directly measure runoff over an entire watershed, but you can measure the watershed's discharge and relate it to the runoff (see sidebars for definitions of these terms).

🖳 Turn on the **Major U.S. Rivers**, **Canadian Rivers**, **Mexican Rivers**, and **Gauging Stations** layers.

The **Gauging Stations** layer shows the locations of gauging stations that measure discharge for each of the six major watersheds. There are actually several gauging stations within each watershed, but for this exercise we are using one located on only the main outflow river of each watershed.

🖳 Click the Identify tool 🛈.

🖳 In the Identify Results window, select the **Major U.S. Watersheds** layer from the list of layers.

🖳 Click in one of the highlighted watersheds to obtain data for that watershed.

8. Record the watershed area (km²) and annual precipitation (cm) for each watershed in columns 3 and 6 of Table 1 on page 84.

🖳 Repeat this process to record the area and annual precipitation for each of the other five watersheds.

🖳 In the Identify Results window, select the **Gauging Stations** layer from the list of layers.

🖳 Click directly on a gauging station symbol to see the data for that station.

9. Record the annual discharge (km³) of the main outflow river for each watershed in column 2 of Table 1.

🖳 Repeat this process to record the annual discharge for each of the other five watersheds.

🖳 Close the Identify Results window when you are finished.

10. For each of the six watersheds, calculate the runoff in cubic kilometers of water per square kilometer of surface area. To do this, divide the value in column 2 by the value in column 3, and record the result (in km) in column 4 of Table 1.

11. Convert the runoff in column 4 from kilometers to centimeters (see sidebar for help with this conversion). Record your results in column 5 of Table 1.

Calculating infiltration and evapotranspiration

In Table 1, you calculated the amount of water that would have to run off every square kilometer of land in each watershed to produce the runoff recorded at the gauging station. Not every drop of precipitation simply flows over land and into a river. Precipitation can also evaporate directly from the ground, infiltrate the soil to be taken up by plant roots and transpired through the leaves, or infiltrate downward, eventually entering the watershed's groundwater.

Evapotranspiration—the combination of evaporation from the soil and from plants through their leaves (transpiration).

Infiltration—the flow of water downward into the soil from the land surface.

The mathematical equation describing runoff is:

Runoff = Precipitation − (Evapotranspiration + Infiltration)

Measuring *evapotranspiration* and *infiltration* are very difficult. Fortunately, you already have information about precipitation and runoff. Therefore, you can determine the amount of evapotranspiration and infiltration by rearranging the above equation. To do this, put all the known quantities (precipitation and runoff, in this case) on the right side of the equation and all the unknown quantities (evapotranspiration and infiltration) on the left side of the equation:

(Evapotranspiration + Infiltration) = Precipitation − Runoff

12. Using the new equation, subtract the runoff (column 5 of Table 1) from the precipitation (column 6 of Table 1) to find the combined Evapotranspiration + Infiltration for each watershed. Record your results in column 7 of Table 1.

13. Use the following equation to calculate the percentage of precipitation that becomes runoff for each watershed and enter it in column 8 of Table 1.

Percent Runoff = (Runoff [column 5] ÷ Precipitation [column 6]) × 100

14. How is the percent runoff for each watershed related to the following:

    a. Watershed size?

    b. Amount of precipitation the watershed receives?

15. How does the percent runoff vary with the climate, shown in column 9 of Table 1? What do you think causes this variation?

16. Which of these major U.S. watersheds do you live in, or is nearest to you?

17. How does the percent runoff you calculated for your watershed affect the following factors in your community:

a. Availability of water?

b. Quality of water?

c. Quantity of water?

🖥 Quit ArcMap and do not save changes.

Table 1— Major U.S. watersheds

1 Watershed name	2 Annual discharge km³	3 Watershed area km²	4 Runoff km³/km² =km	5 Runoff cm	6 Annual precipitation cm	7 Evapotranspiration + Infiltration cm	8 Percent runoff	9 Climate
Where to get data -->	Gauging Stations	Major U.S. Watersheds	Calculate	Calculate	Major U.S. Watersheds	Calculate (precipitation – runoff)	Calculate	
Mississippi River Basin								humid continental to humid subtropical
Nelson River Basin								subarctic
St. Lawrence River Basin								humid continental
Colorado River Basin								arid
Columbia River Basin								semiarid to temperate marine
Rio Grande Basin								semiarid

Wrap-up 2.6

The local water picture

In this unit, you have investigated long-term weather patterns (i.e., climate) and surface runoff patterns across the United States. You have also examined some of the major weather disasters of the last 20 years and considered the ways in which these disasters affect both local and regional communities.

Orographic uplift — the upward movement of an air mass when it encounters mountains. If the air mass contains enough moisture and is cooled sufficiently, precipitation will occur.

Uplift can be triggered by several factors:

- topography
- weather fronts
- winds (jet stream)
- convection (non-frontal thunderstorms)

To conclude this unit, you will examine the precipitation patterns across the United States in the last 24 hours to determine which of the four factors listed above have influenced recent precipitation events. As you will recall, precipitation can occur when uplift causes air to cool.

🖥 Launch ArcMap, then locate and open the **ewr_unit_2.mxd** file.

Refer to the tear-out Quick Reference Sheet located in the Introduction to this module for GIS definitions and instructions on how to perform tasks.

🖥 In the Table of Contents, right-click the **Surface Water Runoff** data frame and choose Activate.

🖥 Expand the **Surface Water Runoff** data frame.

Topography

Topography does not change over a short time, unlike the other three factors listed above. Here, you will consider parts of the country where orographic uplift may play an important role in precipitation.

1. On Map 1 below, outline and shade in areas where precipitation may be caused by orographic uplift.

Map 1 — Orographic uplift and precipitation

Weather Fronts

In Reading 2.3, you learned about weather fronts, or the boundaries between air masses with different temperature and moisture characteristics. Atmospheric uplift and precipitation are often, but not always, associated with weather fronts. Now you will determine the locations of the weather fronts across the country in the last 24 hours, without paying too much attention to the type of front.

🖥 Click the Media Viewer button 🎞.

🖥 Select **Weather Website** and click **OK**.

This will launch a Web browser with the weather Web site automatically. If this does not work, you can open the link in your Web browser at

http://www.rap.ucar.edu/weather

🖥 Click the **Forecast** map.

🖥 Click the **Current Analysis** map (right-hand side) to open a larger version of the map.

On the **Current Analysis** map, warm fronts are identified by red lines and cold fronts are identified by blue lines.

2. Mark the locations of all weather fronts on Map 2 below.

Map 2 — Weather fronts

Weather fronts example

Original map (Feb. 4, 2003)

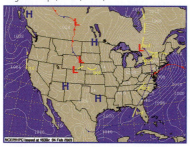

Locations of weather fronts (Feb. 4, 2003)

🖥 When you have finished, click your Web browser's Back button once to return to the Forecast Web page.

Jet Streams

In this unit, you also learned about the position of jet streams and how they are related to precipitation across the U.S.

🖥 On the Forecast Web page, click the **00 hr Forecast** check box and click the **300mb Winds** link.

The map that opens shows the direction and velocity of the jet stream. A scale bar at the bottom of the image shows the wind speeds, with the fastest winds (red and pinks) indicating the location of the jet stream. Remember that there may be more than one jet stream.

3. On Map 3 below, mark the position(s) of the jet stream(s).

Map 3 — Jet streams

Jet Streams example

Original map (Feb. 4, 2003)

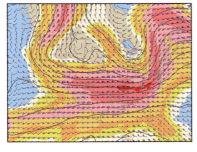

Location of jet stream (Feb. 4, 2003)

When you have finished, click your Web browser's Back button twice to return to the Weather Home page.

Convection (non-frontal thunderstorms)

Another source of atmospheric uplift is convection, the thermal updrafts associated with surface heating. In this unit, you learned that thunderstorm activity is typically associated with convection or with frontal uplift. One defining characteristic of thunderstorms is their great cloud height. The cumulonimbus cloud formations associated with thunderstorms often rise to 18 km (60,000 ft) or higher in the atmosphere. Because of their height, the tops of thunderstorms are often quite cold relative to the tops of other cloud formations lower in the atmosphere. This temperature difference can be seen in images taken by weather satellites that use color-enhanced infrared imagery, in which different colors are used to represent temperature variations in cloud formations.

On the navigation bar of the Weather Home page, click the **Satellite** link.

On the satellite image page, click the **loop** link within the **ConUS ir | loop** link on the right side of the page.

This image may take a few minutes to load, so be patient. The scale bar at the bottom of the image is in degrees Celsius, with red colors denoting temperatures above 0 °C, and greens, blues, and purples representing progressively cooler temperatures below 0 °C. The strongest convection (the area most likely to have thunderstorms or thunderstorm development) is indicated by blues and purples.

Strong convection example

Original map (Feb. 4, 2003)

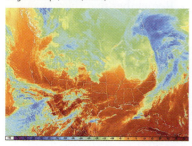

Locations of strong convection (Feb. 4, 2003)

Note that areas of strong convection near fronts have not been outlined.

Precipitation example

Original map (Feb. 4, 2003)

Precipitation (Feb. 4, 2003)

4. On Map 4 below, mark the locations where you see strong convection. Refer to the fronts you drew on Map 2 (page 86) to be sure the convection is not related to the frontal boundaries.

Map 4 — Non-frontal convection

Putting it all together

You have investigated current locations of each of the four factors associated with atmospheric uplift. Next, you will compare these locations to precipitation that has occurred over the past 24 hours.

🖥 Click the Media Viewer button 🎛.

🖥 Select **Intellicast Website** and click **OK**.

This will launch a Web browser with the Intellicast Web site automatically. If this does not work, you can open the link in your Web browser at

http://www.intellicast.com

🖥 Choose **US Weather** under the **US** link at the top of the page.

🖥 Choose **Daily Precipitation** under the **Historic** menu.

5. On Map 5 on the following page, shade areas of the U.S. that received precipitation over the last 24 hours. Create a color or pattern key to identify light and heavy precipitation and indicate these in the precipitation legend at the lower left corner of the map.

6. On the basis of the information you obtained, which of the factors that cause uplift (topography, fronts, jet streams, or convection) appear to be most strongly related to recent precipitation events in the United States? Label them in the appropriate areas on the map above.

Map 5 — Precipitation events in the past 24 hours

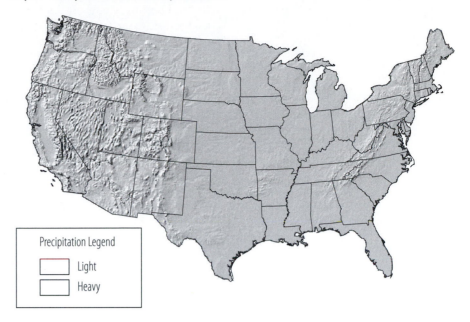

Precipitation Legend

☐ Light

☐ Heavy

7. Which uplift factors played a relatively minor role in the precipitation that occurred over the last 24 hours?

Question 8

Refer to questions 1 and 2 in Investigation 2.4B if you need help answering this question.

8. Were there any jet streams over the U.S. in the images you examined? If so, thinking back to the jet stream movie and the four jet-stream seasonal positions, determine whether the jet stream is where it is supposed to be.

For example...

If you are answering question 9 in January, and you see widespread thunderstorm activity, that is something unexpected in most parts of the U.S. You may need to dig deeper to see if other factors might be influencing recent precipitation.

9. Are the factors responsible for recent precipitation events the ones that you would expect for this time of year? Explain.

🖳 Close the Web browser.

🖳 Quit ArcMap and do not save changes.

Unit 3
Using Water

In this unit, you will

- *Explore the many ways we use water in daily life.*

- *Discover which human activities consume the most water.*

- *Investigate regional patterns in water use and the relationship between water use and population size.*

- *Identify water use classifications and examine the different ways in which water is used and consumed.*

- *Estimate the lifetime of one of the nation's largest aquifers.*

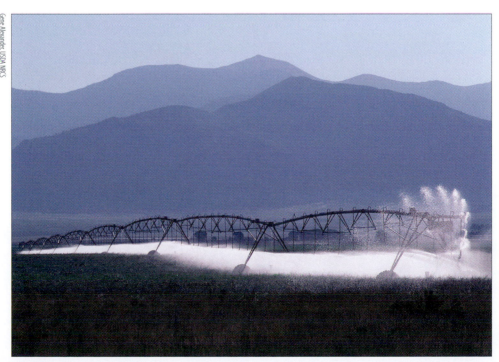

Gene Alexander, USDA NRCS

Center-pivot irrigation system in Colorado.

Warm-up 3.1

Water in your world

Water is one of Earth's most plentiful resources, and one of the most widely used. In earlier units, you explored global water reservoirs and the geographic availability of water for human use. In this unit, you will examine the many ways we use water.

Water at home

Your investigation of water's many uses will begin at home. What are some of the ways in which you use water every day? In addition to direct uses of water, consider how water might be used in making or growing the many products you use.

1. List as many different uses of water in and around your home as you can. (You may attach additional sheets if necessary.)

Water at work and play

Interview five family members, neighbors, or relatives to learn how they use water outside the home, both at work and for recreation. Be prepared to share your lists with your classmates.

2. List as many different uses of water outside the home as you can.

Categorizing water use

Share your list of ways we use water with a small group or your entire class. Then, identify ways to group or classify the types of water use. For example, you might classify the use according to *where* the water is used or *how* the water is used (e.g., household or washing). Provide a brief description of each classification, and be prepared to share these with your class.

3. List and describe your major classifications of water use. (At least five but not more than ten categories.)

Investigation 3.2A

Water for many uses

Earth's most accessible freshwater reservoirs are lakes and rivers on the surface and groundwater below the surface. In Unit 1 you learned that precipitation replenishes these reservoirs as water moves through the hydrologic cycle. In Unit 2 you examined the factors that determine where precipitation falls and what happens to precipitation when it reaches Earth's surface.

In this unit you will examine how we use these freshwater reservoirs in the United States. Water is a renewable resource, but we need to be careful not to extract water from reservoirs faster than it can be replenished through the hydrologic cycle. As you explore these water-use data, watch for patterns that illustrate the challenges we face in managing this important resource.

Who uses the most water?

🖥 Launch ArcMap, then locate and open the **ewr_unit_3.mxd** file.

Refer to the tear-out Quick Reference Sheet located in the Introduction to this module for GIS definitions and instructions on how to perform tasks.

🖥 In the Table of Contents, right-click the **Water Use by State** data frame and choose Activate.

🖥 Expand the **Water Use by State** data frame.

This data frame shows the total amount of water—fresh and saltwater, surface and groundwater—used by each of the 48 contiguous states.

1. Does the total water use in the U.S. show any general geographic patterns? Do northern states use more water than southern states? Do eastern states use more than western states? Large states more than small states?

Where are the data for Alaska and Hawaii?

Water use data for Alaska and Hawaii are not available from government agencies in the same format as for the other 48 states and therefore are not included in this unit.

Capacity versus volume

A *gallon* is a measure of *capacity*, or how much something can hold. A *cubic meter* is a measure of *volume*, or how much space a liquid, for example, occupies. However, even though they are two different measurements, we are comparing them as volumes in this unit for simplicity.

1 cubic meter = 264 gallons

You will examine the data more closely to look for additional patterns.

🖥 Select the **Total water use** layer.

🖥 Click the Open Attribute Table button 🎟 to open the **Total water use** attribute table.

The **Total water use** attribute table shows water-use statistics for each state. Area is given in square kilometers, and water use in millions of gallons per day (mgal/day). First, you will look for general patterns in water use among the states.

🖥 Right-click on the **Total Use (mgal/day)** field heading and choose Sort Descending.

The states are now sorted from highest to lowest total water use.

2. In Column 1 of Table 1, record the six states that use the most water.

Table 1 — Six states with the highest water use, population, and area

States that use the most water	States with the largest population	States with the largest area

💻 Right-click on the **Population** field heading and choose Sort Descending to sort the states by population, in descending order.

3. In Column 2 of Table 1, record the six states with the largest populations.

💻 Right-click on the **Area (km^2)** field heading and choose Sort Descending to sort the states by area, in descending order.

4. In Column 3 of Table 1, record the six states with the largest areas.

5. Based on Table 1, do you think water use is more closely related to a state's population or to its area? Explain.

💻 Right-click on the **Total Use (mgal/day)** field heading and choose Sort Ascending.

The states are now sorted from lowest to highest total water use.

6. In Column 1 of Table 2, record the six states that use the least water, from lowest to highest water use.

Table 2 — Six states with the lowest water use, population, and area

States that use the least water	Population rank	Area rank

7. Using the information in the attribute table, record the population rank and area rank of each state in Table 2. (Note: A rank of 1 represents the largest and 50 represents the smallest).

8. For the six states you listed in Table 2 that use the least amount of water, which factor do you think is more closely related to water use, the population size or the area? Explain.

9. What other factors might explain differences in the amount of water that states use, aside from population and area?

🖥 Close the attribute table.

Water use per capita

Based on what you have seen, it does not seem appropriate to compare the amount of water a state uses without considering the size of the state or the number of users. To adjust for differences in population size, it is best to compare the water use per capita, which is the average amount of water used by each person. The water use per capita is calculated by dividing the amount of water used by the number of users. Use per capita is often expressed in *gpcd* — gallons per capita per day.

Per capita — Latin term that means "for each person."

🖥 Turn off the **Total water use** layer.

🖥 Turn on the **Total use per capita** layer.

🖥 Select the **Total use per capita** layer.

🖥 Click the Open Attribute Table button 🖩 to open the **Total use per capita** attribute table.

🖥 Right-click on the **Per Capita (gpcd)** field heading and choose Sort Descending. Scroll through the table as necessary to complete Table 3.

10. In Table 3, record the names and total water-use rate per capita of the states with the highest and lowest rates of total water use per capita.

🖥 Scroll through the attribute table to find your state.

11. In Table 3, record the name and total water-use rate per capita of the state in which you live.

Table 3 — Highest and lowest per capita total water use

	State	Total water-use rate per capita *gallons/person/day*
Highest		
Your state		
Lowest		

Hint for question 12

Think back to Warm-up 3.1 where you explored some of the *indirect* ways in which you use water.

12. Why might total water use per capita be a poor choice for comparing the amount of water used by people in different states?

🖳 Close the attribute table.

Rather than comparing total water use among states, you can compare the water used just for household or domestic purposes. Domestic use includes water that is used in homes.

🖳 Turn off the **Total use per capita** layer.

🖳 Turn on the **Domestic use per capita** layer.

13. Describe any geographic patterns you see in domestic water use per capita.

🖳 Select the **Domestic use per capita** layer.

🖳 Click the Open Attribute Table button 🔲 to open the **Domestic use per capita** attribute table.

🖳 Right-click on the **Use per capita (gpcd)** field heading and choose Sort Descending to sort the states by their water use per capita

14. In Table 4, record the names and domestic use per capita for the states with the highest and lowest rates of domestic water use, as well as your own state.

Data for domestic use

In rural areas, many people have their own water wells. States generally do not monitor the water withdrawn from private wells. Therefore, the data for domestic and farm use in these areas is underreported. If a home or farm in a rural area receives water from a public supply (surface water or groundwater), that will be reported accurately. In many western states, irrigation water is drawn from public supplies (rivers, etc.), where withdrawals can be measured.

Table 4 — Highest and lowest domestic water use per capita

	State	Domestic water-use rate per capita *gallons/person/day*
Highest		
Your state		
Lowest		

15. What percentage of your state's total water use is domestic use? Percent of Total Use = (Domestic use per capita [Table 4] ÷ Total use per capita [Table 3]) × 100. Round your answer to the nearest tenth of a percent.

16. What percentage of your state's total water use is *not* for domestic purposes? (100% – % domestic use)

17. If the total water use per capita is not all used for domestic purposes, what other ways might this water be used in your state?

🖥 Close the attribute table.

🖥 Quit ArcMap and do not save changes.

Investigation 3.2B

Water for many uses

Water use by sector

This part of the investigation examines water use in greater detail. The government classifies water use into several major groups, or sectors. The major water-use sectors are:

- **Commerce** — water used by commercial facilities and institutions including hotels, restaurants, hospitals, and schools.

- **Domestic** — water used for household purposes.

- **Industry** — water used in the production of steel, chemicals, paper, plastics, minerals, petroleum, and other products.

- **Power** — water used to power steam-driven electric generators. It does not include water used for generating hydroelectric power.

- **Mining** — water used for extracting minerals, oil, and natural gas.

- **Agriculture** — water used to raise animals and irrigate crops.

A simple way to visualize the amount of water used for different purposes is a pie chart. The pie chart below represents the total amount of water used by each sector in the 48 contiguous states (Figure 1). The entire circle represents the total amount of water used for all purposes, and the slices represent the percentage of the total water used in each individual sector.

Water for power

Over 99 percent of the water used for generating electrical power comes from surface water —from both freshwater and saline sources.

Want to know more?

If you would like more information on water used for electrical power, refer to the USGS Web site at:

http://ga.water.usgs.gov/edu/wupt .html

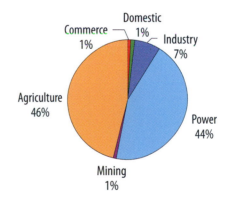

Figure 1. Water-use by sector in the 48 contiguous states.

1. According to Figure 1 above,

    a. which two sectors use most of the water in the U.S.?

    b. what percent of the total water do these two sectors use, combined?

East versus West

🖥 Launch ArcMap, then locate and open the **ewr_unit_3.mxd** file.

Refer to the tear-out Quick Reference Sheet located in the Introduction to this module for GIS definitions and instructions on how to perform tasks.

🖥 In the Table of Contents, right-click the **Water Use by State** data frame and choose Activate.

🖥 Expand the **Water Use by State** data frame.

🖥 Turn off the **Total water use** layer.

🖥 Turn on the **Use by sector** layer.

🖥 Select the **Use by sector** layer.

This layer shows a water-use pie chart for each state. The amount of water used by each state is represented by the size of its pie chart.

2. Examine the legend in the **Use by sector** layer. In the western states, what sector makes up the largest piece of each pie?

3. In the eastern states, what sector makes up the largest piece of each pie?

Next you will compare the amount of water used by the power and agriculture sectors in the eastern and western United States.

🖥 Use the Select Features tool 🔲 to select the 17 western states, as shown below. Do not touch any states east of North Dakota.

How to calculate statistics

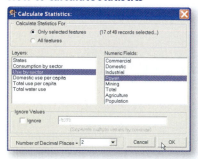

🖥 Click the Statistics button 🔲.

🖥 In the Statistics window, calculate statistics for **only selected features** of the **Use by sector** layer, using the **Power** field.

🖥 Click **OK**. Be patient while the statistics are calculated.

🖥 In the Statistics window, the **Number of Records** should read 17 (you may need to expand the size of the Statistics window to see this). If not, close the Statistics window, reselect the western states, and recalculate the statistics.

The **Total** reported in the Statistics window is the water used for generating power (excluding hydroelectric power) in the western states in millions of gallons per day.

4. Record the amount of water used for power production (**Total**) in the western states in Table 1. Round to the nearest whole number. (Be sure to put it in the column labeled **Water use**, not **Water consumption**).

Table 1 — Water use and consumption in western and eastern states

States	Population	Water use mgal/d		Water consumption mgal/d		Consumption %	
		Agriculture	Power	Agriculture	Power	Agriculture	Power
Western							
Eastern							

- 💻 Close the Statistics window.
- 💻 Click the Statistics button ⓧ.
- 💻 In the Statistics window, calculate statistics for **only selected features** of the **Use by sector** layer, using the **Population** field.
- 💻 Click **OK**. Be patient while the statistics are calculated.

The **Total** reported in the Statistics window is the total population in the western states.

5. Record the total population in the western states in Table 1.

- 💻 Close the Statistics window.
- 💻 Click the Statistics button ⓧ.
- 💻 In the Statistics window, calculate statistics for **only selected features** of the **Use by sector** layer, using the **Agriculture** field.
- 💻 Click **OK**. Be patient while the statistics are calculated.

The **Total** reported in the Statistics window is the total amount of water used for agriculture by the western states.

6. Record the amount of water used for agriculture by the western states in Table 1. Round all values to the nearest whole number.

- 💻 Close the Statistics window.
- 💻 Now select the 31 eastern states by clicking the Switch Selected Features button 🔲. (Note: The whole U.S. may appear highlighted when switching the selection to the eastern states, but it will still calculate the statistics correctly.)
- 💻 Repeat the statistics procedures above to find the population and the amount of water used for power and agriculture in the eastern states.

7. Record the total population and the amount of water used for power and agriculture in the eastern states in Table 1. Round all values to the nearest whole number.

- 💻 Close the Statistics window when you are finished.
- 💻 Click the Clear Selected Features button 🔲.

8. Examine the relationship between population size and the amount of water used by the agriculture and power sectors. What differences do you see between the western states and eastern states?

Use versus consumption

Not all water that is utilized by various water-use sectors is completely *consumed*, or "used up." Often, water used for one purpose can be used again. For example, water used by power plants for cooling can be reused or returned to freshwater reservoirs. Other types of wastewater can also be treated and reused. Water that is returned for reuse is called *return flow*. By subtracting the return flow from the total water used, you can find out how much water has actually been consumed.

💻 Turn on the **Consumption by sector** layer.

This layer also uses pie charts to show the amount of water consumed by the six water-use sectors in each state. The diameter of the pie chart represents the total amount of water consumed in that state.

💻 Leaving the **Use by sector** layer on, turn the **Consumption by sector** layer on and off, comparing the two sets of pie charts.

9. Are the pie charts for water consumption larger or smaller than the pie charts for water use? Why is this?

10. In which of the six water-use sectors do you see the largest difference when you compare water use to water consumption? Explain.

Now you will calculate statistics for water consumption.

💻 Turn off the **Use by sector** layer.

💻 Select the **Consumption by sector** layer.

💻 Use the Select Features tool 📲 to select the 17 western states as you did on page 102.

💻 Click the Statistics button 🗵.

💻 In the Statistics window, calculate statistics for **only selected features** of the **Consumption by sector** layer, using the **Power** field.

💻 Click **OK**. Be patient while the statistics are calculated.

💻 In the Statistics window, the **Number of Records** should read 17, indicating that you have 17 of the 48 states selected. If not, close the Statistics window, reselect the western states, and recalculate the statistics.

The **Total** is the water consumed by power generation in the western states, in millions of gallons per day.

11. Record the amount of water consumed for power production in the western states in Table 1 on page 103. Round to the nearest whole number.

💻 Close the Statistics window.

💻 Click the Statistics button 🅇.

💻 In the Statistics window, calculate statistics for **only selected features** of the **Consumption by sector** layer, using the **Ag Consumed** field.

💻 Click **OK**. Be patient while the statistics are calculated.

The **Total** is the water consumed by agriculture in the western states, in millions of gallons per day.

12. Record the total amount of water consumed by agriculture in the western states in Table 1. Round to the nearest whole number.

💻 Close the Statistics window.

💻 Now select the 31 eastern states by clicking the Switch Selected Features button 🔲. (Note: The whole U.S. may appear highlighted when switching the selection to the eastern states, but it will still calculate the statistics correctly.)

💻 Repeat the Statistics procedures above to find the amount of water consumed by the power and agriculture sectors in the eastern states.

13. Record the amount of water consumed by the power and agriculture sectors in the eastern states in Table 1.

💻 Close the Statistics window when you are finished.

💻 Click the Clear Selected Features button 🔲.

14. Calculate the *percentages* of water consumed by the agriculture and power sectors in both the eastern states and the western states and enter them in Table 1. Round to the nearest percent. Use the formula provided and show your work in the space below before entering values in the table.

% Consumption = (water consumption ÷ water use) × 100

Water consumed in power generation

Water used for cooling in power plants can be reused or returned to freshwater reservoirs, and therefore is not *consumed* or used up. However, there is a lot of water used for other purposes in power generation that is completely consumed and cannot be returned for reuse later.

15. Examine Table 1. Which of the two water-use sectors (agriculture or power) consumes the highest percentage of the water it uses? Explain why you think this occurs.

🖥 Quit ArcMap and do not save changes.

Reading 3.3

Water at work

There is no overall shortage of water on Earth. The continuous movement of water through the hydrologic cycle provides a renewable resource. However, water of sufficient quality and quantity may not be available when and where it is needed. Much of the water we consume is groundwater drawn from underground aquifers that takes tens of thousands of years to accumulate. Using groundwater faster than it is renewed by the hydrologic cycle creates an imbalance. For these reasons, water is considered a finite, or limited, resource; and balancing the many demands for water with the available supply is an ongoing challenge.

How is water used?

To make good decisions about water usage, one has to understand the major usage categories — domestic, commerce, industry, mining, power, and agriculture. You will take a closer look at how water is used in each of these sectors. As you explore these sectors in detail, consider the questions: "How is water being used?" and "How *should* water be used?"

Domestic

Domestic use refers to the water we use in our homes, for things like cooking, drinking, bathing, washing clothes, flushing the toilet, brushing teeth, watering the lawn, and so on.

Water is most often supplied to our homes by a city or county water department or by a private company. This supply is called the public supply. In 1995, about 85 percent of the water used for domestic and commercial purposes came from the public supply. The remainder was self-supplied water from private wells, particularly in rural areas.

Commerce

Commercial water use includes the same activities as domestic use, but applied to commercial businesses (Figure 1). The water required to prepare a meal at a restaurant — to wash produce, mix the syrup for a soda, or clean the floors — is considered commercial use. Water used by the military, schools, prisons, businesses, and hotels is also considered commercial use. Commercial water usually comes from the public supply, although some commercial users have their own wells. Golf courses, for example, are often irrigated with water from private wells.

1. Describe a use of water in the commercial sector in addition to the examples given above.

Commercial water use

Figure 1. Commercial use includes the water used in this self-service car wash.

Industry

In industry, water is often used to manufacture products you own or use. In addition to water being an ingredient in manufactured products, water is also used for washing and cooling machinery employed during the manufacturing process. Industrial water-quality requirements are often less strict than in the domestic and commercial sectors.

Surface water accounts for more than 75 percent of the water used for industrial purposes. This contrasts sharply with the domestic sector, where groundwater is strongly preferred due to consumers' demand for "pure" water. Groundwater is considered pure water because it is filtered through the earth (often from snowmelt), and because it is less likely to be exposed to **contaminants**, or pollutants. Industry is sometimes able to make use of saltwater, which is almost never used in the domestic market.

Industrial water users have an important advantage over other users: Water used for cooling and cleaning can often be reused. High-quality filters remove many of the impurities, permitting the reuse of cleaning water. In cooling systems, water needs only to cool down before it can be used again. Overall, industry consumes only 15 percent of the water it uses—the other 85 percent is returned for reuse.

> 2. Which source supplies the most water to industries—groundwater or surface water? Why?

Mining

In mining, water is used during the extraction of minerals like copper and silver; and of fossil fuels including coal, oil, and natural gas. Water used in stone quarries and ore mills, and to control dust in and around mines is included in mining's water use. Water used in processing minerals and refining fossil fuels is generally reported as industrial use.

Power

Electrical power generators require water for cooling, in much the same way that other industries use water for cooling. Additionally, power is generated by using fuel to heat water to steam, which turns massive turbines that generate electricity. After cooling, this water is either reused or returned to the environment. Nearly all of the water used for power generation is surface water drawn from lakes and rivers.

Water is also used to produce electricity in hydroelectric power plants. Rivers provide the moving water necessary to drive generator turbines. To provide sufficient water pressure, dams are often built, creating artificial lakes. The water stored in these lakes generates a constant supply of power despite seasonal changes in flow, and produces additional power to meet increased demands during the hot summer months. Water in these lakes and reservoirs is also frequently used by nearby farms and cities.

Irrigation methods

Figure 2. *Flood or furrow irrigation.* Where water is relatively cheap, flood irrigation is the simplest and least expensive irrigation method. It requires flat fields to avoid ponding. Flooding is a good way to flush salts out of the soil, but evaporation losses are high.

Figure 3. *Spray irrigation.* More expensive than flood irrigation, spraying is used on crops where flooding is impractical. Evaporation losses are high, and applying water to leaves encourages the growth of plant diseases.

Figure 4. *Drip or microirrigation.* This is the most expensive, but also the most water-efficient irrigation technique. It requires high-quality water (low in salts) to prevent clogging and the buildup of salinity in soil around plants. Drip irrigation is most often used in orchards and vineyards where watering is concentrated on individual trees and the ground is not plowed each year.

3. Why do you think the water used for power generation comes mainly from surface water instead of groundwater?

Agriculture

Water is used in a variety of ways in agriculture, including irrigating crops, raising livestock, and mixing chemicals to apply to crops. This section will focus on the two major uses, crop irrigation and livestock support.

Irrigation

Irrigation uses more water in the United States and around the world than any other activity. Without the ability to supply water to millions of acres of crops, modern agriculture would grind to a halt. The majority of the croplands in the U.S. are located in areas that do not receive reliable precipitation, and must either import surface water from great distances or pump groundwater to irrigate crops. Approximately 37 percent of the water used for irrigation in the U.S. is groundwater, and 50 percent of this water is used by just five states: California, Idaho, Colorado, Texas, and Montana.

Nearly half of the water used for irrigation is lost through leaking pipes and evaporation. Evaporation is the largest source of water loss. Most irrigation involves spraying water into the air over crops. The small water droplets have a large ratio of surface area to volume, resulting in higher evaporation rates compared to large bodies of standing water.

4. Refer to Figures 2, 3, and 4 on irrigation (at left), and suggest which types of crops are best suited to drip irrigation. Explain.

5. What percentage of the water used for irrigation is lost? Why?

Livestock support

In the care and feeding of livestock, drinking water is one obvious use of water. The quantity of water used by farm animals varies greatly depending on the size of the animal. Chickens require very little drinking water (less than a cup per day), whereas cattle drink much more (several gallons per day, though milk-producing cows require even more drinking water than beef cattle).

Water is also used to process food produced from the animals (meat, eggs, milk, and cheese) and to manage animal waste. Management of animal waste creates many environmental problems. This wastewater is typically stored in open-air lagoons. If water from these lagoons enters our freshwater supplies, it poses a serious health risk. During Hurricane Floyd in 1999, many of the freshwater reservoirs used for drinking water in North Carolina were contaminated when lagoons filled with hog waste were flooded by the heavy rainfall and the wastewater was spread throughout the state.

Water sustainability

The overall amount of water used in the United States has remained constant or even decreased slightly since about 1985. In the domestic use sector, water use per capita has decreased due to conservation measures; however, total use has increased by 14 percent due to population growth. Our growing population will eventually overtake our ability to further reduce water use through conservation, and we will be faced with increasingly difficult challenges and decisions. Because water supply and demand are not evenly distributed across the country, some areas will have to deal with more complicated water issues than other areas.

Management of water resources involves both maintaining water quality and determining how water should be used. Because clean, safe drinking water is essential, water quality is of great concern. Cleaning up polluted water supplies is both difficult and expensive. Where contamination has already occurred, cleaning up the mess is our only option. However, it is equally important that we take preventative measures to protect our water supply from further contamination by harmful pollutants.

Issues of water management and sustainability are not unique to the United States. All nations face challenges in providing sufficient amounts of clean, safe water to the public. Global population growth and its relationship to the water supply involve complex, interconnected issues that will figure increasingly in world affairs.

6. Why has domestic water use per capita in the United States decreased, even though the population has increased?

Three Rs of water conservation

There are steps we can take to conserve water. These are sometimes referred to as the three Rs of conservation — **Reduce**, **Reuse**, and **Recycle**. Here are some of the ways in which people save water in each of the major use sectors.

The decline in domestic water use per capita in the United States since 1985 has been a result of conservation efforts such as low-flow toilets and showers, improved home irrigation systems and habits, and effective water-conservation education programs. According to studies by the Environmental Protection Agency, the average family of four now uses 20,000 fewer gallons a year than they did a few years ago. That is enough water to fill a small backyard swimming pool.

The commercial, industrial, and mining sectors have reduced water use significantly in recent years. New laws and financial incentives have encouraged businesses to develop more efficient processes and practices that use less water, and use more recycled water. Car washes and factories collect and filter wastewater and reuse it many times before it is discharged into the sewer system.

Although electric power generating plants use tremendous quantities of water for cooling, virtually all of the water withdrawn is returned or reused many times. Most power plants are built near surface water sources, and are capable of using low-quality water that is not suitable for domestic or agricultural use. In fact, almost one-third of the water used in generating electricity is saltwater.

Recall that roughly half of the water used to irrigate crops is wasted. The waste can be reduced by using alternative irrigation methods (Figures 2, 3, and 4). Drip-irrigation systems use one third as much water as traditional spray-irrigation systems, because they irrigate plants directly rather than by spraying water into the air. Drip irrigation is not practical or appropriate for many crops, but it is very effective in orchards and vineyards that are not plowed and replanted each year.

Another step suggested by some conservationists is to reduce the quantity of meat in our diets, thus reducing the quantity of water used in the production of livestock. Water is used to grow feed, clean up waste, and process livestock into food. It has been estimated that it takes as much as 45,360 liters (approximately 12,000 gallons) of water to produce a pound of beef, compared to 416 liters (approximately 110 gallons) to produce a pound of wheat. Such a suggestion would certainly not be supported by the livestock industry or people who enjoy meat products. Clearly, it is a complex issue.

7. Describe three ways in which business and industry help to conserve freshwater.

Investigation 3.4

Feeding a nation

In an earlier investigation, you learned about six major water-use sectors —domestic, commerce, industry, mining, power, and agriculture. Of these, agriculture consumes by far the most water. In this investigation, you will examine patterns in agricultural water use and discuss the implications of those patterns.

Precipitation and irrigation

🖥 Launch ArcMap, then locate and open the **ewr_unit_3.mxd** file.

Refer to the tear-out Quick Reference Sheet located in the Introduction to this module for GIS definitions and instructions on how to perform tasks.

🖥 In the Table of Contents, right-click the **Water for Agriculture** data frame and choose Activate.

🖥 Expand the **Water for Agriculture** data frame.

🖥 Scroll down the Table of Contents to read the precipitation legend.

This data frame shows the average annual precipitation for each county of the contiguous United States, in centimeters per year. The climate of a region is determined in part by the amount of precipitation it receives annually. Major climate types and the annual precipitation amounts associated with each are shown in Table 1.

Food exports

In addition to producing food for our own consumption, U.S. farmers and ranchers export many agricultural products to other countries. In 2005, these products accounted for $85 billion, or 9.2 percent of our total exports.

Reading the precipitation legend

To see the **Precipitation** layer legend, scroll to the bottom of the Table of Contents.

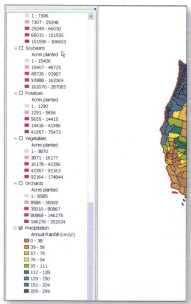

Table 1 — Major climate types

Climate type	Precipitation	
	cm/yr	in/yr
Arid	under 30	under 12
Semiarid	30 – 50	12 – 20
Sub-Humid	50 – 75	20 – 30
Humid	75 – 200	30 – 80
Tropical Humid	more than 200	more than 80

1. Based on precipitation, which climate type would you assign to the western U.S.? Which climate type would you assign to the eastern U.S.?

   a. Western U.S.

   b. Eastern U.S.

The climate difference between the eastern and western halves of the country has shaped the development of agriculture in the U.S. The East has adequate rainfall to support all but the most water-thirsty crops. In the West, precipitation is unreliable, and sufficient amounts of surface water are not available. Therefore, large-scale agriculture in the West depends on groundwater to irrigate crops.

🖥 Turn on the **Water for irrigation (% total)** layer.

This layer shows the percentage of each county's total water consumption that is used for irrigation. Notice that some counties use nearly all of their water for irrigation.

🖥 Turn the **Water for irrigation (% total)** layer off and on several times, and look for patterns related to precipitation.

You can see that areas with low precipitation depend heavily on irrigation. However, if you look closely you will also notice regions that are extensively irrigated, even though those regions receive significant rainfall (over 125 cm) each year.

2. On Map 1 below, indicate three regions that use large amounts of water for irrigation, yet also receive more than 125 cm of precipitation per year.

Map 1 — Highly irrigated regions with high precipitation

Missing county data

Some counties either did not collect or did not supply water data, or were allowed to withhold their data to protect the economic interests of their farmers.

Sorghum — a crop that is a member of the grass family, much like wheat and corn. It is used widely in the cereal, snack food, baking, and brewing industries.

This data frame includes layers that show where major crops are grown. Next, you will compare individual crop layers to the irrigation layer to find out which crops are grown in regions where the natural precipitation is supplemented by irrigation.

🖥 Turn off the **Water for irrigation (% total)** and **Precipitation** layers.

🖥 Turn each individual crop layer on and off, beginning with **Corn** and ending with **Orchards**. Examine the regions you identified on Map 1 to see if the crop is grown in that region.

3. On Map 1, label each region you identified in question 2 with the names of the crops grown in that region.

Subsidized irrigation water

The U.S. Bureau of Reclamation provides about one quarter of the irrigation water in the West at wholesale prices. Water quantities and prices are set by long-term contracts, and are highly subsidized by taxpayers. Farmers are charged an annual fee according to the acreage of crops served, rather than the actual amount of water used. Subsidies average $54 per acre, but may be as high as $150 per acre.

How big is an acre?

An acre is 43,560 square feet, or equal to about 0.4 hectare in the metric system. It is roughly the size of a football field.

Missing county data

Some counties either did not collect or did not supply water data, or were allowed to withhold their data to protect the economic interests of their farmers.

Agricultural water sources

In some areas, there is enough water in nearby lakes and rivers to meet irrigation needs. The cost to pump and deliver this local surface water ranges from free to about $15 per acre per year. In the dry climate of the West, the Bureau of Reclamation has built dams, reservoirs, and canals to capture and deliver water to farms. This more distant surface water costs $10 to $90 per acre, but is heavily *subsidized*, or financially supported, by the government. In areas with expensive or unreliable surface-water supplies, farmers may pump groundwater. Due to the high cost of pumping, groundwater costs from $11 to $105 per acre. Many farmers use a combination of ground- and surface water to meet their irrigation needs.

🖳 Turn off everything except the **States** layer.

🖳 Turn on the **Primary irrigation source** layer.

This layer represents the primary source of water used for irrigation. Counties colored brown use groundwater, and counties colored blue use surface water as the primary source for irrigation.

4. On Map 2 below, mark at least three large regions east of the Rocky Mountains that depend on groundwater for irrigation.

Map 2 — Regions dependent on groundwater for irrigation in eastern U.S.

Aquifer — a water-bearing rock formation.

🖳 Turn on the **U.S. aquifers** layer.

🖳 Click the Identify tool ⓘ .

🖳 In the Identify Results window, select the **U.S. aquifers** layer from the drop-down list of layers.

🖳 Click on the aquifers to determine the names of the aquifers that provide groundwater for the regions you marked on Map 2.

5. Label each of the regions you marked on Map 2 with the name of the major aquifer(s) that provides irrigation water for that region.

🖳 Close the Identify Results window.

🖳 Turn off the **U.S. aquifers** and **Primary irrigation source** layers.

The High Plains Aquifer

The largest of these aquifers, the High Plains Aquifer — also called the Ogallala Aquifer — covers some 450,000 square kilometers (174,000 square miles) in Texas, Oklahoma, Nebraska, New Mexico, Colorado, Kansas, and South Dakota.

💻 Turn on the **High Plains Aquifer** layer.

💻 Turn on the **Crops by county (% acreage)** layer.

The **Crops by county (% acreage)** layer displays small pie charts showing the percentage of farm acreage used for nine major crops. You can get a good idea of the major crops grown in different regions by looking at the color patterns formed by the pie charts.

💻 Use the Zoom In tool 🔍 to zoom in on the High Plains Aquifer. (Hint: Turn the **Crops by county (% acreage)** layer off and on to see the location of the High Plains Aquifer under the pie charts.)

6. What are four major crops irrigated by the High Plains Aquifer?

   a.

   b.

   c.

   d.

Crop coefficients (Kc)

In this chart, bermuda grass is the standard crop, with a Kc value of 1. A lower crop coefficient number indicates a lower water need compared to bermuda grass.

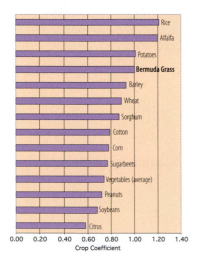

Agricultural scientists have developed a measure of the water needs of different crops. This measurement is known as the "crop coefficient" (Kc). This measurement compares the amount of water needed by each crop to a standard crop such as grass or alfalfa under the same environmental or climate conditions. Farmers use crop coefficients, along with average temperature, latitude, and other climate factors, to determine their irrigation needs.

7. According to the crop coefficient chart at left, how do the water needs of the crops you listed in question 6 compare with the water needs of other crops?

💻 Turn off the **Crops by county (% acreage)** layer.

💻 Click the Full Extent button 🌐 to view the entire map.

Squeezing the sponge

The High Plains Aquifer is composed of layers of sand, clay, and gravel eroded from the Rocky Mountains. Water is stored in the sand and gravel layers. The thickness of the water-bearing sediments of the aquifer ranges from less than a meter to several hundred meters.

When farmers first tapped this water source in the late 1930s, they thought it would provide an unlimited supply of water. This is probably one reason why farmers grew high-water-use crops in that region. However, the overlying rock is fairly *impermeable*, which means that it allows very little water to infiltrate the ground and reach the aquifer. Consequently, the aquifer's recharge rate—the rate at which water is replenished from the surface—is very low. In fact, the recharge rate averages only about 0.9 cm (0.35 inches) per year across the aquifer. Next, you will calculate how long the aquifer can support agriculture at the current rate of groundwater withdrawal.

- 🖥 Turn on the **Primary irrigation source** layer.

- 🖥 Click the Select By Location button 🖳.

- 🖥 In the Select By Location window, construct the query statement:

    I want to **select features from** the **Primary irrigation source** layer that **have their center in** the features of the **High Plains Aquifer** layer.

- 🖥 Click **Apply**.

- 🖥 Close the Select By Location window.

This will highlight the counties within the boundaries of the High Plains Aquifer. Next, you will determine how much groundwater these counties use for irrigation.

- 🖥 Click the Statistics button 🗙.

- 🖥 In the Statistics window, calculate statistics for **only selected features** of the **Primary irrigation source** layer, using the **Groundwater (mgal/d)** field.

- 🖥 Click **OK**. Be patient while the statistics are calculated.

The **Total** is the total amount of groundwater withdrawn in the counties served by the aquifer, in millions of gallons per day (mgal/d).

8. How many million gallons of groundwater are withdrawn from the aquifer each day?

How to calculate statistics

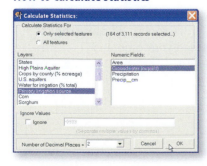

Converting from gallons to cubic meters

1 cubic meter = 264 gallons

Therefore,

millions of gallons ÷ 264 g/m³
= millions of cubic meters.

9. Convert the withdrawal rate to cubic meters per year.

   a. Multiply by 1,000,000 to convert from millions of gallons per day to gallons per day. (Hint: Just add six zeroes.)

   b. Divide by 264 gal/m³ to convert from gallons per day to cubic meters per day. Round the result to the nearest 100,000 m³/d.

   c. Multiply by 365 days/year to find the annual withdrawal from the aquifer in cubic meters per year. Round to the nearest 100,000 m³/yr.

Capacity versus volume

A *gallon* is a measure of *capacity*, or how much something can hold. A *cubic meter* is a measure of *volume*, or how much space a liquid, for example, occupies. However, even though they are two different measurements, we are comparing them both as volumes in this unit for simplicity.

The total surface area of the aquifer is 450,000,000,000 m² (450 billion square meters) and the recharge rate is 0.009 m/yr.

10. Calculate the total annual recharge to the aquifer in m³/yr by multiplying the area of the aquifer by the recharge rate.

11. What is the difference between the annual rate of withdrawal (from question 9c) and the rate of recharge of the High Plains Aquifer (from question 10)? Is this an annual gain or loss of water for the aquifer?

The total volume of available water in the High Plains Aquifer is estimated at 4.03 trillion m³ (4,030,000,000,000 m³).

12. For how many more years will the High Plains Aquifer last at the current rates of recharge and withdrawal? (Hint: Divide the total volume of the aquifer by the annual loss you calculated in question 11.)

13. Do you think this estimate is realistic? What factors could cause the High Plains Aquifer to run out of water sooner than your estimate?

Recall that much of the water we consume is groundwater that can take tens of thousands of years to accumulate in aquifers. Using groundwater faster than it is renewed by the hydrologic cycle creates an imbalance. For these reasons, water is considered a *finite*, or limited, resource.

14. What do you think could realistically be done to protect this immense but finite resource in the High Plains Aquifer? Propose a series of actions and a time line for implementing those actions. For example, what could you change about the type and acreage of the crops grown there, or the irrigation practices?

🖳 Quit ArcMap and do not save changes.

Feeding a nation

Wrap-up 3.5

Meeting the challenge

One of the goals of this unit is to examine patterns of water use and our ability to sustain that use. Here are three primary issues in water management.

- **Quantity** — Is there enough fresh, clean water to meet the various needs of the population?

- **Proximity** — Are the sources of water close to where it is used, or far away?

- **Quality** — Is the water clean and pure enough for its intended use?

Discuss these water challenges as a class or in small groups, and list any additional challenges that you identify. You may wish to re-examine the water-use data using ArcMap to explore these issues in more detail.

1. List water challenges that your state, county, or city are facing, or will probably face in the near future.

2. What are the financial, political, and environmental challenges your state may face as a result of these water issues listed above? Give specific examples from the water-use data you have explored in the investigation, from current events, or from personal knowledge and experience.

3. How do these issues affect you personally? How might they change your
    lifestyle or behavior?

4. If you were in charge of water policy for your state, what measures would
    you recommend, in the face of these challenges, to ensure an adequate
    water supply for the next 50 years?

Unit 4
Water for a Desert City

In this unit, you will

- *Identify water-related challenges faced by desert and western cities.*

- *Examine the importance of maintaining a water balance.*

- *Determine the economic and environmental consequences of excessive groundwater pumping.*

- *Investigate water-use patterns and the importance of water conservation.*

- *Develop a plan to help Tucson, Arizona meet its future water needs.*

Downtown Tucson, Arizona and the Catalina Mountains.

Warm-up 4.1

The four great deserts of the Southwest

Figure 1. Red dots represent cities in the four major deserts of the southwestern U.S. and Mexico.

Hint for question 1

Refer back to Warm-up 1.1 and Investigation 1. 2 in Unit 1 to refresh your memory of all the water reservoirs on Earth.

Living in a desert

Desert regions are characterized by low annual precipitation and a high evaporation rate. These regions, covering approximately 20 percent of the land on Earth, may be large expanses of shifting sand with few plants, or rugged mountainous regions colonized by many plant species.

In the United States, the four large deserts of the Southwest—the Chihuahuan, Great Basin, Mojave, and Sonoran deserts—contain many cities with significant populations (Figure 1). One of these cities is Tucson, the second largest metropolitan area in Arizona, with a population of over 800,000. This region was originally occupied by Native Americans and later by Spanish settlers who established missions and military posts beginning in 1638. After Arizona became part of the United States in 1912, the population of Tucson grew as people realized the potential for mining and agriculture. The continued population growth of the Tucson area has had important consequences for the environment. The demand for more water is met by drilling new wells and importing surface water from distant sources.

1. Which kind of water reservoir is tapped by pumping water from wells?

2. Speculate on the environmental consequences of removing water from this reservoir faster than it is replenished by the hydrologic cycle.

3. Do you think that issues of water supply and demand are unique to desert cities and towns? Explain how the problems of obtaining adequate water for desert inhabitants might apply to your city or town.

The Santa Cruz River was a critical water source for early Tucson inhabitants. Archaeologists have found evidence of agricultural settlements along the Santa Cruz dating back to approximately 1000 B.C. Compare the two photographs on page 127 taken of the west branch of the Santa Cruz River. Both photographs were taken from the same location, 96 years apart.

4. Describe the major differences in the environment between the two photographs.

5. What do you think might have caused the changes?

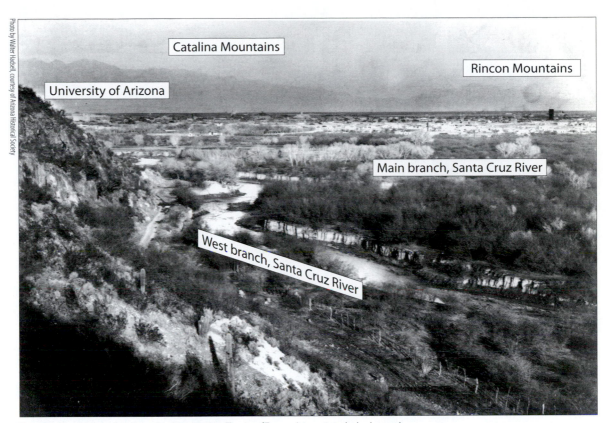

Figure 2. West branch of the Santa Cruz River in 1904. The city of Tucson, Arizona is in the background.

Figure 3. View from the same location in 2000. The west branch of the Santa Cruz River has been filled in, but the main channel is still visible, above center.

Investigation 4.2A

Aquifer — a water-bearing rock formation.

Drainage basin — an area in which runoff flows downhill to a common point, such as a lake; or to a common channel, such as a river or stream.

Water in the balance

Maintaining a balance between the water we use and the water supply that is replenished by precipitation is the key to ensuring that cities and agriculture continue to thrive. This is a particularly demanding task in desert regions. The deserts of the American Southwest contain bustling cities and an extensive network of farms and ranches that depend on water for survival.

Tucson, Arizona is a desert city that relies on groundwater withdrawn from the underlying aquifer to meet almost all its water needs. Therefore, to maintain a water balance, we must understand when, where, and how the water in an aquifer is replenished. In this investigation, you will examine the factors critical to replacing water in an aquifer.

Precipitation patterns in the Tucson Basin

Precipitation events provide most of the water that replenishes aquifers. Therefore, examining precipitation patterns is the first step to understanding aquifer recharge.

🖥 Launch ArcMap, then locate and open the **ewr_unit_4.mxd** file.

Refer to the tear-out Quick Reference Sheet located in the Introduction to this module for GIS definitions and instructions on how to perform tasks.

🖥 In the Table of Contents, right-click the **Precipitation Patterns** data frame and choose Activate.

🖥 Expand the **Precipitation Patterns** data frame.

This data frame shows a shaded relief image of the Tucson area as well as the city streets.

🖥 Turn on the **Precipitation Zones** layer.

The colors that are displayed in this layer correspond to the amount of annual precipitation received by different areas, averaged over a 30-year period. The heavy dashed line marks the approximate boundary of the Tucson Basin, including the slopes of the mountains that drain into the basin.

1. Use the legend to determine the highest precipitation range in the Tucson Basin (the area inside the heavy dashed line).

2. What color represents the highest precipitation range found in the Tucson Basin?

3. Is the amount of annual precipitation the same across the Tucson Basin? Explain your observations.

🖥 Turn on the **Weather Stations** layer.

This layer shows the location of weather stations within and outside the Tucson Basin.

🖥 Click the Identify tool 🛈.

🖥 In the Identify Results window, select the **Weather Stations** layer from the drop-down list of layers.

🖥 Using the Identify tool 🛈, determine which weather stations inside the Tucson Basin (inside the dotted line on the map) receive the highest and the lowest amounts of annual precipitation. You may need to scroll down to see the annual precipitation inside the Identify Results window.

4. In Table 1, record the location of weather stations that receive the highest and lowest amount of annual precipitation in the Tucson Basin. Include the amount of annual precipitation they receive and the month they receive the most precipitation.

Table 1 — Highest and lowest annual precipitation in the Tucson Basin

	Location of station	Annual precipitation *cm/yr*	Month of highest precipitation
Station with the highest precipitation			
Station with the lowest precipitation			

5. Do the data you entered in Table 1 above support your answer to question 3? Explain.

🖥 Close the Identify Results window.

🖥 Turn off the **Precipitation Zones** and **Weather Stations** layers.

Aquifer recharge

Precipitation is continually replacing, or *recharging* groundwater that people remove from the aquifer. Of course, not all precipitation ends up in the aquifer. As you learned in previous activities, precipitation is diverted through different parts of the hydrologic cycle when it reaches land.

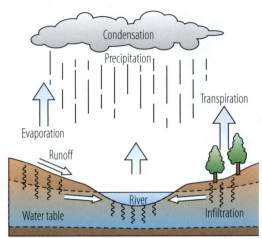

Figure 1. Land processes of the hydrologic cycle.

Examine the hydrologic cycle diagram above to explore what can happen to precipitation when it reaches the ground.

6. Which processes in the hydrologic cycle recharge the aquifer?

Natural recharge

Precipitation infiltrates basin soils and, to a lesser extent, joints, fractures, and faults in the bedrock of the surrounding mountain ranges. If precipitation falls where the ground is too wet or where water is unable to penetrate the surface, that water becomes runoff and joins streams and rivers. In the Tucson Basin, streams and rivers are usually dry because the *water table*, or the top boundary of the aquifer, lies below the river bottoms (Figure 1). After precipitation events, these water channels fill with runoff and merge together as they flow downstream and eventually drain into the Santa Cruz River.

Stream flow

The V shape, which forms where streams merge, points downstream.

- 🖥 Turn off the **Streets** layer.
- 🖥 Turn on the **Streams** layer.

This layer illustrates the network of streams and rivers that flow through the Tucson Basin. The V shape that is formed where two streams merge always points downstream, as shown at left.

7. In which direction does water flow out of the Tucson Basin? (North is "up.")

Gauging station

One type of gauging station is called a stilling well, illustrated below.

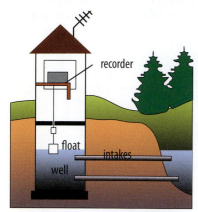

Stream water enters the well through intake pipes. The water level in the well is the same as the water level in the stream. As the water level rises or falls, the float in the well also rises or falls. A cable attached to the float drives a device that records the water level or transmits it to a satellite for recording at a central location (adapted from USGS).

🖥 Turn on the **Gauging Stations** layer.

A river or stream's rate of flow, usually expressed in m³/sec (or ft³/sec), is called its *discharge*. Discharge is measured by instruments at *gauging stations*. This layer shows the locations of gauging stations in the Tucson Basin. Data from these stations will help you determine whether discharge increases or decreases as water moves downstream.

🖥 Click the Identify tool ⓘ. (Note: You may need to click the Pointer tool ▸ before clicking the Identify tool, in order to open the Identify Results window again.)

🖥 In the Identify Results window, select the **Gauging Stations** layer from the drop-down list of layers.

🖥 Use the Identify tool ⓘ to find the name of each gauging station.

8. Label the gauging stations in the boxes of Diagram 1 below.

Diagram 1 — Rillito Creek gauging stations

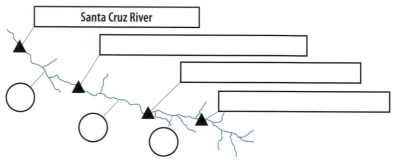

9. In the circles, indicate the number of streams that enter Rillito Creek between gauging stations.

10. Do you expect the discharge to increase or decrease as the water flows out of the Tucson Basin? Explain your answer.

Spiky data

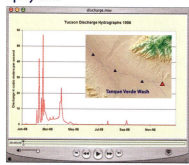

The data in the **Discharge Hydrograph movie** appear more "spiky" than the precipitation data you looked at earlier, because they are daily measurements rather than monthly averages. Furthermore, the gauging station data are for a single year (1998) rather than a 30-year average.

🖥 Close the Identify Results window.

🖥 Click the Media Viewer button 🎞 and choose the **Discharge Hydrograph movie**.

This movie shows the amount of discharge flowing through the four gauging stations on the Rillito Creek and Santa Cruz River as surface waters leave the Tucson Basin.

🖥 Watch the movie several times to determine whether discharge increases or decreases between each of the stations.

11. Look at the discharge between the months of February and May. How does the discharge change between the following gauging stations? (Does it increase, decrease, or remain constant?)

    a. Tanque Verde Creek and Rillito Creek at Dodge Blvd.

    b. Rillito Creek at Dodge Blvd. and La Cholla Blvd.

12. Compare your observations in question 11 to your prediction in question 10. Explain any differences between your prediction and observations.

Watch a flash flood!

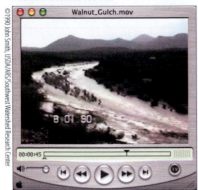

To view a movie of a flood caused by summer thunderstorm precipitation runoff, click the Media Viewer button 🎞 and choose **Flash Flood movie**.

You may have noticed that there is a small, but measurable discharge at the Santa Cruz River gauging station throughout the year, as indicated by the blue dashed line. This line represents *base flow*, which is the result of pumping treated wastewater from the sewer system into the river channel.

13. What is the approximate base flow, in m^3/sec, at the Santa Cruz River station?

14. What might be the benefit of adding this water to the river system?

💻 Close the Media Viewer window.

💻 Turn off the **Streams** and **Gauging Stations** layers.

Estimating annual recharge

Next you will estimate the mean amount of water recharged each year in the Tucson Basin aquifer.

💻 Turn on the **Recharge Regions** layer.

The **Recharge Regions** layer shows the three main recharge regions of the Tucson Basin. Only the portion of the mountain slopes that drain into Tucson Basin are included.

💻 Click the Identify tool ⓘ. (Note: You may need to click the Pointer tool ▸ before clicking the Identify tool, in order to open the Identify Results window again.)

💻 In the Identify Results window, select the **Recharge Regions** layer from the drop-down list of layers.

💻 Using the Identify tool ⓘ, click in each recharge region to gather information to record in Table 2 on the following page.

15. Record the mean annual precipitation and the area for each recharge region in Table 2 on the following page.

Table 2 — Calculating recharge in the Tucson Basin

Recharge area name	Mean precipitation *m* *click in region using the Identify tool* ⓘ	Area *m²* *click in region using the Identify tool* ⓘ	Precipitation volume *m³* *= area x mean precipitation*	Recharge volume *m³* *= precipitation volume x 0.05*
Tucson Mountains				
Central Basin				
Catalina – Rincon Mountains				
Total *(add values in each column)*				

16. Calculate the volume of precipitation that falls in each region by multiplying the mean precipitation by the area. Record your results in the Precipitation volume column of Table 2.

Research has shown that only about 5 percent of the precipitation that falls in the Tucson Basin recharges the aquifer.

17. Calculate the volume of water recharged in each region of the Tucson Basin by multiplying the precipitation volume by 0.05 (5%) and record your results in the Recharge volume column of Table 2.

18. Calculate and record the Total Area, Precipitation volume, and Recharge volume by adding the values in each column of Table 2.

19. Based on what you know about the hydrologic cycle and the climate of the southwestern U.S., why do you think the recharge rate (5%) is so low in the Tucson Basin?

The Colorado River — a desert lifeline

Originating in the Rocky Mountains, the Colorado River is a lifeline for the southwestern United States. Most of the territory the river crosses on its way to the Gulf of California is desert. Along the way, water is withdrawn to meet the needs of 20 million people and to irrigate millions of acres of farmland.

Acre-foot — the amount of water needed to cover one acre of land to a depth of 1 foot. One acre-foot equals about 1230 cubic meters (or 325,851 gallons of water), the amount used by an average family of four in one year.

🖳 Close the Identify Results window.

Artificial recharge

So far you have explored natural recharge of water to the aquifer by precipitation. The aquifer can also be *artificially* recharged with treated wastewater and by surface water imported from other areas. For example, the Central Arizona Project (CAP), completed in 1993 at a total cost of $4 billion, delivers approximately 1.9 billion cubic meters (1.5 million acre-feet) of Colorado River water to cities in central and southern Arizona each year. The CAP canal carries open surface water along a 538-km (336-mi) trip through the desert from Lake Havasu on the Colorado River to a point just north of Tucson, where it is recharged into the aquifer. Pumping stations and pipelines provide some help along the way to get across areas of higher elevation.

🖳 Collapse the **Precipitation Patterns** data frame.

🖳 Right-click the **Supply and Demand** data frame and choose Activate.

🖥 Expand the **Supply and Demand** data frame.

This data frame shows Arizona counties, the CAP canal system, and selected cities that receive CAP water.

🖥 Click the Identify tool 🛈.

🖥 In the Identify Results window, select the **Arizona Cities** layer from the drop-down list of layers.

🖥 Using the Identify tool 🛈, click on Tucson (the southernmost city) to gather information about the city.

Allotment — the volume of water a city, state, or country (U.S. and Mexico) is assigned each year from the Colorado River.

20. What is the annual allotment of CAP water delivered to Tucson in cubic meters (m^3)?

21. Using Table 2 on page 134, compare Tucson's annual CAP allotment with the annual recharge of the Tucson Basin.

22. How do you think Tucson and the Tucson Basin Aquifer would be affected if Tucson's CAP allotment were reduced or cut off completely?

🖥 Close the Identify Results window.

How is Tucson's water supply used?

Like a bank account, maintaining a water balance requires that the water removed from the aquifer be replaced. It is important, then, to keep track of who is using water and how much is being used. In this section you will look at how water is used in Pima County, of which Tucson is the largest city.

Water is used for many different purposes beyond drinking and typical household uses.

23. List examples of how water is used in each of the following water-use classifications.

    a. Public water supply.

Public water supply — includes water used in homes and businesses, as well as public use in parks and recreational facilities.

    b. Mining.

c. Agriculture.

d. Industrial.

24. Speculate on the percentage of groundwater designated for use in the public water supply, mining, agriculture, and industry. Draw a pie chart showing your estimates. Be sure to label each segment.

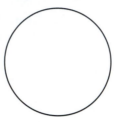

🖥 Turn off the **CAP Canal** layer.

🖥 Turn on the **Water Usage** layer.

This layer displays pie charts that show the breakdown of water usage for each county in the state.

25. Draw and label the actual pie chart for groundwater usage in Pima County.

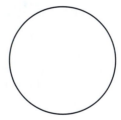

26. How did your estimate compare to the actual chart? Which part of your estimate was most in error?

🖥 Quit ArcMap and do not save changes.

Investigation 4.2B

Water in the balance

In addition to understanding when, where, and how the water in an aquifer is recharged, we must also understand how water is used. In this investigation, you will examine classifications of water use and how factors such as population growth and climate conditions affect water availability in Tucson, Arizona.

Tucson's water balance sheet

As mentioned earlier, assessing Tucson's water situation is similar to maintaining a bank account. There are deposits: recharge from precipitation, treated wastewater, and CAP water added into the aquifer. There are also withdrawals from the aquifer: water used by the people and businesses of Tucson. The goal, like balancing a checkbook, is to make sure that there are enough water deposits into the aquifer to replace the amount of water withdrawn from the aquifer.

In the Tucson Water Balance Sheet, you will balance the "water checkbook" to see if there are enough deposits of surface water to match the withdrawals of groundwater from the aquifer, based on the city's population in 2000 and projected into the year 2050.

💻 Launch ArcMap, then locate and open the **ewr_unit_4.mxd** file.

Refer to the tear-out Quick Reference Sheet located in the Introduction to this module for GIS definitions and instructions on how to perform tasks.

💻 In the Table of Contents, right-click the **Supply and Demand** data frame and choose Activate.

💻 Expand the **Supply and Demand** data frame.

💻 Click the Media Viewer button 🔬 and choose **Tucson Water Balance Sheet**. Or, you can locate the **Tucson Water Balance Sheet** Excel file in the **Unit 4** folder and open it in Microsoft Excel.

Per capita — per person. For example, water use per capita describes the amount of water each person would use if Tucson's total water use was divided equally among everyone living in Tucson.

This spreadsheet allows you to adjust the rate of population growth, the rate of increase in water usage (per capita per year), and the amount of mean precipitation per year. By adjusting these variables, you can find combinations of these factors that will result in a water balance.

Examine the Water Balance graph.

1. If growth and water usage (including the use of CAP water) in Tucson continues at the current rate, will Tucson ever achieve a water balance (as indicated by positive values on the graph labeled "Water Balance")?

2. During which year will Tucson come closest to achieving a water balance?

Deficit — a negative balance, or the amount needed to bring the balance back up to zero.

Hint for question 3

On many newer computers, you can use the Pointer Tool to find the exact value of a bar in the bar graph by holding it anywhere over that particular bar. A small window will appear with the exact value.

☐ Practice adjusting each of the variables to get a feel for how each will affect the water balance.

☐ Set each variable back to its **mean** value when you are done.

3. In the year 2010, what would be the deficit in water availability (in m^3) if the population growth rate increased from the current 2.4 percent to 5 percent

   a. with CAP water?

   b. without CAP water?

4. With a 0.1 percent rate of increase in water usage and a mean annual precipitation of 30 cm, what would be the maximum population growth rate that Tucson could support (with CAP water) and still achieve a water balance in 2050?

5. With a 0.3 percent rate of increase in water usage and a mean annual precipitation of 30 cm, what would be the maximum population growth rate Tucson could support (with CAP water) to achieve a water balance in 2050?

6. Which variable (population growth rate, rate of increase in water use, or precipitation) do you think has the least influence on water balance? Explain.

Scientists have predicted a trend of hotter summers and drier winters in the future.

7. If Tucson's annual precipitation drops to 21 cm per year, what combination of population growth rate and rate of increase in water usage would the city require in order to achieve a water balance with CAP water in 2050? Find 3 combinations and enter them in Table 1 below.

Table 1 — Maintaining a water balance with decreased precipitation

Combination	Population growth rate %	Water-use annual increase %	Annual precipitation cm
1			21
2			21
3			21

8. Are the combinations of population growth rate and increase in rate of water usage that you entered in Table 1 reasonable? Explain.

9. List three ways that citizens or politicians could promote a lower rate of population growth to help Tucson achieve a water balance by 2050.

   a.

   b.

   c.

10. Which of the three ideas listed above is most reasonable? Why?

🖳 Close the Excel spreadsheet. Do not save your work.

🖳 Quit ArcMap and do not save changes.

Reading 4.3

The Tucson Basin aquifer

Aquifer basics

An aquifer is a formation of rock or sediment that is *saturated*, or filled to capacity, with water. Knowing the geology of an aquifer and its overlying sediments is important to understanding water **infiltration** and recharge.

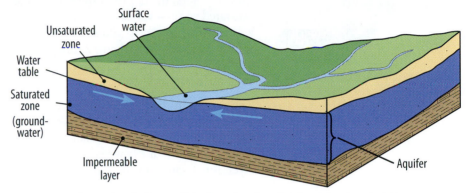

Figure 1. Structure of a typical aquifer. Rivers and lakes form where the surface drops below the water table.

The ability of water to infiltrate, or percolate down through the sediments to reach the aquifer is dictated by **permeability**. If an aquifer is permeable, water can move into it from overlying sediments. The ability of water to be stored in an aquifer is determined by **porosity**. If an aquifer is porous, it possesses empty spaces capable of retaining water. As shown in Figure 2, the quality of an aquifer is determined by both the permeability and porosity of the rock formation. Valuable aquifers are both permeable and porous, allowing water to flow through empty spaces that are connected within the rock.

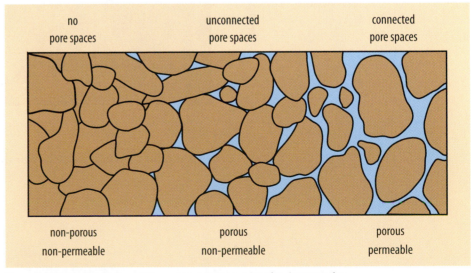

Figure 2. High porosity and permeability are characteristic properties of productive aquifers.

The Basin and Range aquifers

Tucson is located in the Basin and Range Province, a geographic region that extends from southern Idaho into Sonora, Mexico; and from eastern California to central Utah (Figure 3). This region is characterized by mountain ranges that run north-south and are separated by low basins. Viewed from the air, the Basin and Range region has been described as looking like "an army of caterpillars marching toward Mexico."

Sediments eroded from the surrounding mountains fill the basins, and can be from a few hundred to more than 3000 m thick. Over millions of years, water has infiltrated these sediment-filled basins, forming the Basin and Range aquifers. In these aquifers, the depth of the water table—the upper boundary of the aquifer's saturated zone—varies considerably, ranging from the surface (in flowing rivers and streams) to nearly 400 m below the surface. The Basin and Range aquifer system includes more than 72 independent watersheds covering over 200,000 km^2. These aquifers have played a critical role in the growth and development of the southwestern U.S.

The aquifer in the Tucson Basin is composed primarily of sedimentary rock including sand and clay. The cross-sectional diagram of the basin below shows the composition of the sedimentary rock that fills the basin and the historic and current location of the water table. In 1940, the water table was located in the sand layer, much closer to the surface. Since then, it has dropped more than 50 m.

Basin and Range aquifers

Figure 3. Tucson, Arizona is located within the Basin and Range Province. The Basin and Range network of aquifers are shown here in blue.

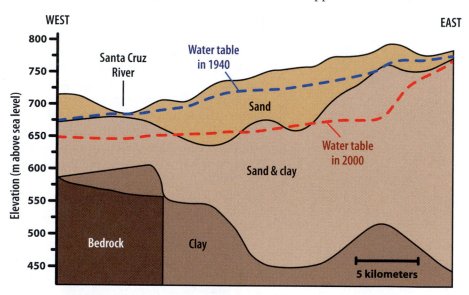

Figure 4. Cross section of the Tucson Basin aquifer showing changes in the water table from 1940 to 2000.

1. What type of rock is associated with the Tucson Basin Aquifer?

Recharging the aquifer

Water removed from an aquifer is replaced or recharged primarily by precipitation. However, most of this precipitation is consumed by other hydrologic-cycle processes and does not recharge the aquifer. For example, much of the water becomes runoff, flows into drainages, and is carried away from the area. Because Tucson's water table is tens to hundreds of meters below the surface in some places, streams and rivers that form the drainage network are dry except during periods of heavy precipitation. Water is also lost through evaporation from the soil and absorption and transpiration by plants. Only 5 percent of the precipitation falling in the Tucson Basin eventually recharges the aquifer, although the process can take quite a long time. Scientists studying recharge of the Tucson Basin aquifer have found that precipitation from the 1960s still has not infiltrated down to the main aquifer.

Colorado River water

In 2001, the Tucson Water Department completed construction of a facility where Colorado River water delivered by the Central Arizona Project (CAP) canal can recharge the underlying aquifer. Once the water infiltrates the surface soil and recharges the aquifer, it can be pumped out of the aquifer and delivered to the city of Tucson. Eleven large ponds, or basins, each ranging from approximately 80,000 – 162,000 m² (20 – 40 acres) in area are filled with CAP water. This water infiltrates the soil under these basins in order to reach the aquifer underneath. Since 1997, over 160 million m³ (130,000 acre-feet, or over 42 billion gal) of water have recharged the underlying aquifer, and over 98 million m³ (80,000 acre-feet, or over 26 billion gal) of the newly recharged water have been pumped out of the aquifer and delivered to Tucson since 2001. Water quality is constantly monitored in both the delivered CAP water and water that has already recharged the aquifer. Since the new recharge facility was constructed and started delivering recharged CAP water, the need to pump groundwater from the Tucson Basin has eased and several wells in the area have been put on standby status, meaning that they are not currently being used but are available to pump if necessary.

A. K. Huth, The SAGUARO Project

Figure 5. Recharge basins filled with CAP water in the desert west of Tucson. The basins allow CAP water to infiltrate the soil and recharge the underlying aquifer.

2. What is the principal source of recharge in the Basin and Range aquifers?

3. How has recharging with Colorado River water changed Tucson's need for pumping groundwater from the Tucson Basin?

Keeping up with demand

It is important to balance the removal of water from the aquifer and the rate at which the aquifer is being naturally or artificially recharged. There are severe environmental consequences to withdrawing more water than is recharged. When the water table drops, the aquifer may be compressed by the weight of the overlying sediments, a process called **compaction**. Compaction closes the empty pore spaces that previously held water, reducing the porosity and permeability of the aquifer, thus slowing infiltration and reducing the aquifer's ability to store water. Once this happens, the aquifer's porosity and permeability cannot be restored.

Compaction of the aquifer due to groundwater pumping also results in **subsidence**, or the lowering of the ground surface. Subsidence can cause considerable damage to sewer, water, and gas pipes as well as buildings and roads. Sewer pipes rely on gravity to maintain flow, so a small change in slope due to subsidence can result in sewage backflow—a serious problem! Intensive groundwater pumping can also drive up water prices and decrease water quality. As existing wells are deepened and new wells are drilled to reach the lower water table, the cost of these upgrades are passed on to the consumer. More power is required to lift the groundwater, raising energy costs. Water retrieved from deeper in the Earth is also likely to be of lower quality because salinity increases with depth.

4. What problems can result from removing more groundwater from an aquifer than the amount replaced by recharge?

Investigation 4.4

Groundwater issues

Imagine coming home and finding cracks in your walls or, worse, that your house is actually *sinking*. You and your home could be one of the many victims of the over-pumping of groundwater. In this investigation, you will determine the extent and severity of the effects of over-pumping in the Tucson Basin.

The physical effects of subsidence

🖳 Launch ArcMap, then locate and open the **ewr_unit_4.mxd** file.

Refer to the tear-out Quick Reference Sheet located in the Introduction to this module for GIS definitions and instructions on how to perform tasks.

🖳 In the Table of Contents, right-click the **Physical Impacts of Subsidence** data frame and choose Activate.

🖳 Expand the **Physical Impacts of Subsidence** data frame.

As you learned earlier, an aquifer is a water-bearing layer of rock in which water fills the spaces, or pores, between sediment particles. The water helps strengthen the rock against the pressure of the overlying material. As water is pumped out of the aquifer, the weight of the overlying material collapses the pore spaces and squeezes the sediment particles together. As a result of this compaction,

- The porosity and permeability of the aquifer irreversibly decrease, reducing the aquifer's storage capacity.
- The ground surface gradually sinks, a process called *subsidence*.

The Compaction Movie

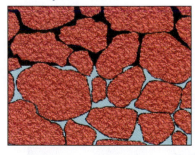

🖳 Click the Media Viewer button 🖳 and choose the **Compaction Movie**.

This animation simulates what happens when groundwater is removed from an aquifer and the water table drops. Look for evidence of the reduction in porosity and permeability, and of compaction.

1. What would happen if you tried to recharge the aquifer again after subsidence had occurred? Would you be able to fill it with as much water as was there originally?

🖳 After viewing the movie several times, close the Media Viewer window.

Using satellite-based radar, scientists precisely measured the elevation of the Tucson Basin floor in 1993 and again in 1997. Subtracting the 1997 elevation data from the 1993 data produced the image you see in the data frame on the computer, called a *radar interferogram*. It shows the elevation changes that occurred over this 4-year period.

The elevation changes are color-coded, and appear as *interference fringes*, similar to the rainbows you see on the surface of a soap bubble due to tiny differences in the

Radar interferometry

This simulated cross section through the center of the Tucson Basin (dashed blue line in the top image) shows recent subsidence. Vertical distances are exaggerated.

thickness of the bubble wall. In the interferogram, each cycle of colored fringes —from one blue fringe to the next blue fringe, for example—equals 2.8 cm of elevation change. The fringes were then overlaid on a satellite photo in order to see city landmarks underneath the interferogram.

💻 Click the Media Viewer button 🎦 and choose the **Interferogram Movie**.

The **Interferogram Movie** shows a 3-D model of the ground subsidence in the Tucson Basin. Elevation changes in the movie are greatly exaggerated for emphasis.

💻 View the movie several times, then close the Media Viewer window.

Next, you will look at a cross-sectional view of the Tucson Basin.

💻 Select the **Cross section** layer.

💻 Using the Hyperlink tool ⚡, click on the blue cross-section line.

The window that appears shows a cross section of the Tucson Basin aquifer and the location of the water table in 1940 and 2000.

2. Describe the change in the water-table elevation from 1940 to 2000.

3. According to the cross section, where has the greatest change in the water table elevation occurred relative to the river and higher elevations?

4. If the groundwater withdrawal continues at the same rate as it has since 1940, draw and label your prediction of the depth of the water table in 2050 on Diagram 1.

Diagram 1 — Water-table elevation

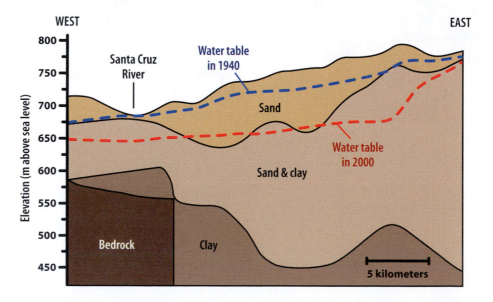

💻 Close the image window.

💻 Turn off the **Cross section** layer.

💻 Examine the **Radar Interferogram** layer.

5. How many interference fringes do you see? (Count the rainbow-colored bands.)

💻 Turn on the **Radar Surface Change** layer.

This layer summarizes the subsidence data shown in the radar interferogram, with each interference fringe shown as a different color.

💻 Click the Identify tool **ⓘ**.

💻 In the Identify Results window, select the **Radar Surface Change** layer from the drop-down list of layers.

💻 Using the Identify tool **ⓘ**, click in each colored subsidence region to collect data about that region. Read the maximum subsidence and area from the Identify Results window.

6. For each subsidence region, record the maximum subsidence that occurred and the area of the region in Table 1 below. Round the area to the nearest whole number.

Table 1 — Subsidence in the Tucson Basin

Band	Maximum subsidence cm	Area km²	Rate of subsidence cm/yr	% of city affected
outer				
middle				
inner				

💻 Close the Identify Results window.

Calculating rate of subsidence

Rate of subsidence (cm/yr) =

Maximum subsidence (cm) ÷ 4 years

Calculating percent of city

Percent area = (area [km²] ÷ 500 km²) × 100

7. Calculate the rate of subsidence (to the nearest tenth of a centimeter per year) for each band over the 4-year period from 1993–1997 and record it in Table 1.

8. If the city of Tucson covers an area of about 500 km², calculate the percentage of the city covered by each band (to the nearest tenth of a percent) and record it in Table 1.

9. If the subsidence rate remains constant, how much subsidence will occur in the center of the basin (innermost band) between 1997 and 2025?

    a. Calculate the number of years between 1997 and 2025.

    b. Then multiply your answer from 9a by the subsidence rate (cm/yr) of the inner band of the city.

The economic impact of subsidence

In addition to the physical impacts of over-pumping groundwater, there are economic impacts as well. In this section, you will estimate the cost of repairing damage to the sewer system caused by subsidence.

💻 Collapse the **Physical Impacts of Subsidence** data frame.

💻 Right-click the **Economic Impacts of Subsidence** data frame and choose Activate.

💻 Expand the **Economic Impacts of Subsidence** data frame.

The subsidence you calculated assumed that the population and water use rate in the Tucson Basin will remain the same as it is today. This layer shows the subsidence predicted by the U.S. Geological Survey for the Tucson Basin between now and the year 2025, based on expected changes in population and water use.

💻 Turn on the **Sewer Damage** layer.

💻 Select the **Sewer Damage** layer.

The **Sewer Damage** layer shows the locations of main sewer lines within the predicted subsidence area. Gravity controls the flow in sewer lines, so even small elevation changes can cause sewer lines to break, stop flowing, or even flow backward!

💻 Click the Statistics button ⊠.

💻 In the Statistics window, calculate statistics for **all features** of the **Sewer Damage** layer, using the **Length (m)** field.

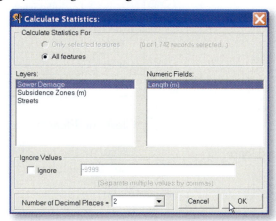

💻 Click **OK**. Be patient while the statistics are calculated.

The **Total** is the total length of sewer mains, in meters, within the subsiding area.

10. What is the total length of sewer mains within the subsiding area? Round your answer to the nearest whole number.

💻 Close the Statistics window.

When the ground in the center of the basin subsides 0.3 meters, about 80 percent of the sewer mains need to be replaced. Next, you will determine how much it might cost the city to replace these sewer mains. To simplify your calculations, assume that subsidence across the basin between now and 2025 is 1.5 meters.

11. Determine the length of sewer mains that must be replaced each time the ground subsides 0.3 meters by multiplying the total length of sewer mains in the subsiding area (from question 10) by 80% (0.80). Round your answer to the nearest whole number.

12. Determine the number of times the sewer mains will need to be replaced by 2025. (Hint: Divide the total amount of predicted subsidence in 2025 (1.5 meters) by 0.3 meters (the amount of subsidence that will require sewer main replacement).)

13. Determine the total length of sewer mains that must be replaced by 2025 by multiplying the number of replacements (from question 12) by the total length of sewer mains that would require replacement (from question 11).

According to a local engineering firm, the cost of replacing a sewer main is $200/meter.

14. How much will it cost the city to replace damaged sewer mains between now and 2025?

Roads, sewers, and utilities are part of a city's *infrastructure*, the basic facilities needed for a city to function.

15. In addition to the sewer mains, what other parts of Tucson's infrastructure do you think are affected by ground subsidence?

16. Describe how each of these infrastructure issues could affect each of the following groups:

   a. Tucson Basin residents.

   b. Residents of the surrounding area.

   c. Seasonal residents and vacationers in Tucson.

   💻 Quit ArcMap and do not save changes.

Investigation 4.5

Conserving water

Water rates increase over time for several reasons. Rates must be raised to cover the costs of operating, maintaining, and expanding the water system, and paying employee salaries. Water rates may also be increased to encourage customers to conserve water. The reasoning is that, as water rates increase, customers will decrease their water use in order to avoid higher monthly bills. In this section, you will investigate how residential water rates have changed through time in Tucson. You will also determine how much water *really* costs.

Inflation — an increase in the prices of products and services over time. Inflation can also be described as a decrease in the purchasing power of money over time. By "adjusting for inflation," economists can realistically compare prices at different times.

Tucson's water rates

The figures below show the population and water rates in the Tucson metropolitan area over several decades. The water rates in Figure 2 have been adjusted for inflation to dollar amounts in the year 2000.

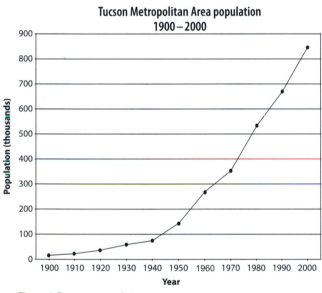

Figure 1. Tucson area population.

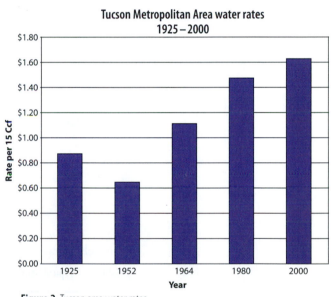

Figure 2. Tucson area water rates.

What does "Ccf" mean?

C = centrum (Latin: one hundred)

Ccf = 100 cubic feet

= 2.83 m³

= 748 gallons

1. According to Figure 1, approximately how large was the Tucson Metropolitan Area population in 1925? How large was the population in 2000?

a. 1925.

b. 2000.

2. Calculate the percent increase in population using the following formula:

$$\% \text{ increase} = \frac{2000 \text{ population} - 1925 \text{ population}}{1925 \text{ population}} \times 100$$

3. According to Figure 2 (previous page), what was the water rate for a 15 Ccf monthly usage in the Tucson Metropolitan Area in 1925? What was the rate in 2000?

   a. 1925.

   b. 2000.

4. Calculate the percent increase in the 15 Ccf-per-month water rate from 1925 to 2000 using the following formula:

$$\% \text{ increase} = \frac{2000 \text{ rate} - 1925 \text{ rate}}{1925 \text{ rate}} \times 100$$

5. Is the percent increase in water rate more than, less than, or similar to the rate of population increase? Why do you think this is?

Table 1 shows the amount of water used by a family of four to do common household tasks.

Table 1 — Typical household water usage for a family of four

Task	Gallons per day	Days per month	Gallons per month
Showering	80	30	80 x 30 = 2400
Flushing toilet	100	30	
Brushing teeth	8	30	
Washing dishes	15	30	
Cooking & drinking	12	30	
Irrigation	48	30	
Kitchen sink use	5	30	
Laundry (1 load)	35	17	
		Total gallons per month	

6. Calculate the monthly amount of water used by a family of four for each of these tasks and record it in the last column of Table 1. The first one has been done for you.

7. Add the number of gallons per month used for all tasks in Table 1, and record the total at the bottom of the last column.

8. Convert the total number of gallons per month to hundreds of cubic feet (Ccf) per month. (Hint: 1 Ccf = 748 gallons.)

9. Using the value you calculated for total Ccf per month in question 8, and the water rates you identified in question 3, calculate what a monthly water bill would have been in the years 1925 and 2000. Use the following formula:

$$\text{monthly water bill for year} = \text{Ccf per month} \times \text{water rate for year}$$

   a. 1925.

   b. 2000.

10. How much did the monthly water bill increase from 1925 to 2000?

11. Calculate the cost per gallon of water in 2000 using the following formula.

$$\text{Year 2000 cost per gallon} = \frac{\text{Year 2000 monthly bill (question 9b)}}{\text{Total gallons per month (Table 1)}}$$

12. If bottled water costs $1.00 per gallon, how many times more expensive is bottled water compared to tap water? (Hint: Divide $1.00 by your answer to question 11.)

Imagine what it would cost to meet all of your family's water needs with *bottled* water!

13. How do you think doubling water rates would affect the city's population and the city's water supply? Explain.

14. Do you think increasing water rates is a good way to encourage people to conserve water? Explain.

Stop the pumping!

You have investigated subsidence and its potential for damaging homes, water mains, and sewer lines. In theory, the solution to many of these problems is easy —stop pumping groundwater from subsiding areas. However, what effect would this have on Tucson's water supply?

🖥 Launch ArcMap, then locate and open the **ewr_unit_4.mxd** file.

Refer to the tear-out Quick Reference Sheet located in the Introduction to this module for GIS definitions and instructions on how to perform tasks.

🖥 In the Table of Contents, right-click the **Wells** data frame and choose Activate.

🖥 Expand the **Wells** data frame.

This data frame shows the major streets in the city of Tucson. The pink area represents the area where significant ground subsidence is occurring.

🖥 Turn on the **Wells** layer.

This layer shows the locations of drinking-water wells in the Tucson Active Management Area.

Active Management Areas

The Tucson Active Management Area (TAMA) is one of five active management areas in Arizona, designed to protect groundwater resources within those areas. The boundary of TAMA is roughly the boundary of the groundwater basin.

🖥 Click the Select By Location button 🗺.

🖥 In the Select By Location window, construct the query statement:

I want to **select features from** the **Wells** layer that **intersect** the features in the **Radar Subsidence** layer.

🖥 Click **Apply**.

🖥 Close the Select By Location window.

The drinking-water wells within the subsidence zone will be highlighted.

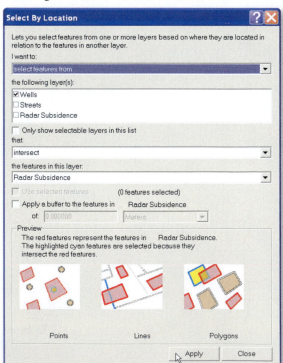

⌨ Click the Statistics button ⊠.

⌨ In the Statistics window, calculate statistics for **only selected features** of the **Wells** layer, using the **Production (gal/year)** field.

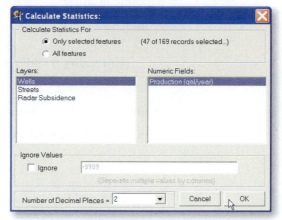

⌨ Click **OK**. Be patient while the statistics are calculated.

In the Statistics window, the number of wells in the subsidence area is the **Number of Records**, and the total annual production (gal/yr) of these wells is the **Total**. (Hint: You may need to increase the size of the Statistics window to see all the results.)

15. How many drinking-water production wells (**Number of Records**) are located within the central Tucson subsidence area?

16. If the wells within the subsiding area were removed from production, how many gallons (the **Total**) would no longer be available for human use?

17. How do you think the city might make up for this shortfall?

⌨ Close the Statistics window.

⌨ Click the Clear Selected Features button ⊠.

Water conservation

Research has shown that only about 5 percent of the precipitation that falls in the Tucson Basin is recharged into the aquifer. A significant amount of precipitation runs off into storm drains and sewer systems. In this section, you will calculate how much water can be *harvested*, or captured and used, from a small area during a single summer storm.

⌨ Collapse the **Wells** data frame.

⌨ Right-click the **Summer Harvest** data frame and choose Activate.

⌨ Expand the **Summer Harvest** data frame.

This data frame shows an aerial photo of the University of Arizona campus, located in central Tucson. The blue features are University buildings. The **University Buildings** layer contains information about the area covered by each building, in square meters.

🖳 Click the Statistics button ☒.

🖳 In the Statistics window, calculate statistics for **all features** of the **University Buildings** layer, using the **Area (m^2)** field.

🖳 Click **OK**. Be patient while the statistics are calculated.

In the Statistics window, the total area covered by the buildings is the **Total**.

18. What is the total area covered by campus buildings, in square meters?

19. Following the directions below, calculate how much water could be collected from the rooftops of the buildings on campus if the storm produced 2.5 cm (1 in) of rain.

   a. Convert centimeters to meters of rain (100 cm = 1 m).

   b. Calculate the volume of rain on the rooftops, in cubic meters. (Hint: Volume = area of rooftops [m²] calculated in question 18 × depth of water [m] calculated in question 19a.)

   c. Convert the volume of water collected from the rooftops to gallons. (Hint: Multiply your answer from question 19b by 264 gal/m³.)

Watch a summer thunderstorm

Storm_081701.mov

00:00:33

To view time-lapse movies of summer thunderstorms, click the Media Viewer button 🎬 and choose **Thunderstorm Movie 1**, **2**, or **3**. (Don't spend too much time watching the storms — you still have work to do.)

USDA/ARS/Southwest Watershed Research Center

20. How could water be collected from the rooftops of the buildings in the main campus area?

21. For what purposes do you think this harvested water could be used?

22. How does the volume of water that can be harvested from campus rooftops (question 19c) compare to the mean annual well production in the subsidence zone (question 16)? Write your answer in terms of a percentage of annual well production. (Hint: Divide the answer to question 19c by the answer to question 16 and multiply the result by 100.)

23. If the roof area of an average-sized home is 150 m², how much water could be collected from the roof of an average home in the same storm?

    a. Convert 2.5 cm of rainfall to meters.

    b. Multiply your answer to 23a by the roof area to get the volume of water in m³.

    c. Convert your answer to 23b to gallons.

Hint for question 23c:

Remember that 1 cubic meter (m³) is equal to approximately 264 gallons.

24. Referring to Table 1, determine what percentage of the average family's monthly irrigation needs could be met using the water harvested from the rooftop in this storm. (Hint: Divide the answer to question 23c by the number of gallons used for monthly irrigation [Table 1] and multiply the result by 100.)

25. Do you think this is a reasonable technique for conserving water in Tucson? What issues might complicate this type of conservation?

🖳 Quit ArcMap and do not save changes.

Wrap-up 4.6

The voice of conservation

Voicing your ideas on conservation

Now that you have examined the environmental, physical, and economic issues surrounding Tucson's water situation, you may have developed some of your own solutions to the water supply problem. This section provides you with an opportunity to express your ideas and develop a conservation plan for the city of Tucson, using the knowledge you gained in this investigation.

Project

You have been hired by Tucson Water as a community water conservation specialist. Your job is to communicate with citizens about ways to conserve water and the importance of doing so. Write a letter that will be mailed to all residents, explaining why they should care about conserving water and what they can do to save water. In this letter, describe three different action plans by which residents can help decrease groundwater use in Tucson. Give specific facts and details about your plans, about what will happen if residents do not conserve, and about the benefits they will enjoy if they do conserve. You might want to refer to values you calculated in Investigation 4.2B (in the Tucson Water Balance Sheet) and Investigation 4.5.